Cambridge IGCSE®

Design & Technology

STUDENT'S BOOK

Also for Cambridge O Level

Justin Harris, Dawne Bell, Chris Hughes, Matt McLain, Stewart Ross, David Wooff

William Collins' dream of knowledge for all began with the publication of his first book in 1819. A self-educated mill worker, he not only enriched millions of lives, but also founded a flourishing publishing house. Today, staying true to this spirit, Collins books are packed with inspiration, innovation and practical expertise. They place you at the centre of a world of possibility and give you exactly what you need to explore it.

Collins. Freedom to teach

Published by Collins
An imprint of HarperCollins*Publishers*
The News Building
1 London Bridge Street
London SE1 9GF

Browse the complete Collins catalogue at
www.collins.co.uk

© HarperCollins*Publishers* Limited 2018

10 9 8 7 6 5 4 3 2 1

ISBN 978-0-00-829327-7

Justin Harris, Dawne Bell, Chris Hughes, Matt McLain, Stewart Ross and David Wooff assert their moral rights to be identified as the authors of this work.

All rights reserved. No part of this publication may be reproduced, stored in a retrieval system, or transmitted in any form or by any means, electronic, mechanical, photocopying, recording or otherwise, without the prior written permission of the Publisher or a licence permitting restricted copying in the United Kingdom issued by the Copyright Licensing Agency Ltd., Barnard's Inn, 86 Fetter Lane, London, EC4A 1EN.

British Library Cataloguing in Publication Data
A Catalogue record for this publication is available from the British Library

Commissioned by Lisa Todd

Project edited by Rachel Nickolds

Project managed by Ben Gardiner

Edited by Lucy Hyde

Proofread by Helen Davies and Cassandra Fox

Indexed by Malcolm Henley

Illustrations by Stewart Ross, Terry Bream and Jouve India Private Limited

Cover design by Gordon MacGilp

Cover illustrated by Maria Herbert-Liew

Typeset by Jouve India Private Limited

Production by Rachel Weaver

Printed and bound by Grafica Veneta

With thanks to our reviewers: Terry Bream (UK) and John Zobrist (Singapore)

Thanks also to the students of Jerudong International School, Brunei, for contributing examples of their coursework.

IGCSE® is the registered trademark of Cambridge Assessment International Education.
Exam-style questions and sample answers are written by the authors.

MIX
Paper from
responsible sources
FSC™ C007454

This book is produced from independently certified FSC paper to ensure responsible forest management.

For more information visit:
www.harpercollins.co.uk/green

Contents

Introduction 4

Section 1
Product Design 6
1·1 Getting started 8
1·2 Design ideas and techniques 20
1·3 Making 36
1·4 Evaluation 48
1·5 Health and safety 55
1·6 Use of technology 63
1·7 Design & Technology in society 72
1·8 Product design application 78
1·9 Environment and sustainability 101

Section 2
Graphic Products 108
2·1 Formal drawing techniques 110
2·2 Sectional views, exploded drawings and assembly drawings 115
2·3 Freehand drawing 120
2·4 Drawing basic shapes 122
2·5 Developments 127
2·6 Enlarging and reducing 130
2·7 Instruments and drafting aids 134
2·8 Layout and planning 136
2·9 Presentation 139
2·10 Data graphics 144
2·11 Reprographics 148
2·12 Materials and modelling 150
2·13 ICT 156
2·14 Manufacture of graphic products 159

Section 3
Resistant Materials 164
3·1 Types of materials 166
3·2 Smart and modern materials 170
3·3 Plastics 173
3·4 Wood 177
3·5 Composites 184
3·6 Metals 188
3·7 Preparation of materials 195
3·8 Setting and marking out 199
3·9 Shaping 204
3·10 Joining and assembly 210
3·11 Finishes 222

Section 4
Systems and Control 228
4·1 Systems 230
4·2 Structures 233
   4·2·1 Basic concepts 233
   4·2·2 Types of frame structure members ...237
   4·2·3 Strengthening frame structures 240
   4·2·4 Nature of structural members 242
   4·2·5 Applied loads and reactions 245
   4·2·6 Moments 248
   4·2·7 Materials 251
   4·2·8 Testing 253
   4·2·9 Joints in structures 255
   4·2·10 Forces 258
4·3 Mechanisms 265
   4·3·1 Basic concepts 265
   4·3·2 Conversion of motion 273
   4·3·3 Transmission of motion 281
   4·3·4 Energy 292
   4·3·5 Bearings and lubrication 295
4·4 Electronics 299
   4·4·1 Basic concepts 299
   4·4·2 Circuit building techniques 305
   4·4·3 Switches 311
   4·4·4 Resistors 316
   4·4·5 Transistors 320
   4·4·6 Diodes 323
   4·4·7 Transducers 326
   4·4·8 Capacitors 328
   4·4·9 Time delay circuits 330
   4·4·10 Logic gates and operational amplifiers 335

Section 5
The Project 344

Glossary 359
Index 366
Acknowledgements 373

Introduction

Design & Technology has many applications for everyday life. For example, the design process requires problem solving skills that can be applied to any problem you encounter. Design & Technology is a very 'real world' subject – everything manmade around you has been designed, from the chairs you sit on to the pens and pencils you write with. As you learn about each of the topics in this book, try to relate them to the products, materials and systems you see and use on a daily basis. Looking closely at familiar, everyday products in this way will not only help you to develop your Design & Technology skills, but will also make the course more interesting and enjoyable.

After working through the contents of this book, you will have a good understanding the principles of Product Design. You will also look at either: Graphic Products, Resistant Materials or Systems and Control in more depth as a specialism. You will learn about specific materials, their properties, how they are made, and the processes used to manipulate their shapes and size, in order to turn them into useful designs.

The book is organised into the following sections:

Section 1: Product Design	This section examines the role of the designer, the design process, health and safety, design and the environment, design application and design in society.
Section 2: Graphic Products	Here you will learn how to communicate your ideas effectively. You will develop a number of techniques, from improving your formal drawing skills and understanding the importance of modeling prototypes, to using Computer Aided Design (CAD) and ICT to develop and present ideas quickly and easily.
Section 3: Resistant Materials	This looks at a broad range of materials, techniques, tools, equipment, methods and processes. This section will help you to develop your understanding of both the physical and working properties of a range of woods, metals and plastics.
Section 4: Systems and Control	This section looks closely at structures, mechanisms and electronics, and how these technological areas can be linked together to design controllable systems.
Section 5: The Project	This focusses on a designing and making activity, based on one of the specialisms above. This will be submitted as your coursework. The project allows you to explore materials, be creative, apply critical thinking, analyse and produce an outcome to solve a problem.

This book has some very useful features that have been designed to help you understand all the aspects of Design & Technology that you will need to know as you work through the course. These are shown on the next page.

Throughout the course you will be required to demonstrate your knowledge and understanding of the theories and techniques, facts, terms and concepts included in the book. You will also need to demonstrate an understanding of the wider application of Design & Technology, plus an ability to analyse and evaluate products in terms of their design and production.

We hope you enjoy the course and that you find this book a useful and valuable companion on your Design & Technology journey of discovery. Good luck!

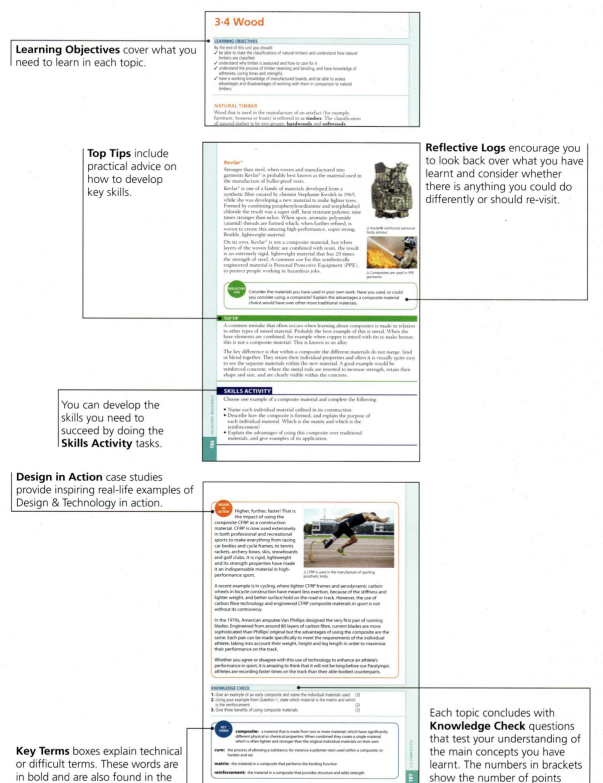

Learning Objectives cover what you need to learn in each topic.

Top Tips include practical advice on how to develop key skills.

Reflective Logs encourage you to look back over what you have learnt and consider whether there is anything you could do differently or should re-visit.

You can develop the skills you need to succeed by doing the **Skills Activity** tasks.

Design in Action case studies provide inspiring real-life examples of Design & Technology in action.

Key Terms boxes explain technical or difficult terms. These words are in bold and are also found in the Glossary at the back of the book.

Each topic concludes with **Knowledge Check** questions that test your understanding of the main concepts you have learnt. The numbers in brackets show the number of points available for each question.

Wherever you are in the world, you are surrounded by items that are designed and created by people.

Product design is a process of bringing together a number of disciplines, such as science, art, mathematics, engineering and technology, to make products that can be sold to customers. Product design is broadly concerned with creating, innovating and developing new, and hopefully useful, products.

Designers generally follow a systematic process when creating and innovating. Initially they will receive a brief. They must first research this brief, and fully understand the problem they are working on and who the new product is for. Once this task is complete they can start the interesting and creative element of the process, which is to come up with new ideas. These ideas need to be evaluated continually until they are ready to be made into models, prototypes and, eventually, result in manufacture, distribution and sales.

Understanding the design process, and how to take a problem and turn it into a product that is marketable, is an interesting and challenging task. Knowledge of this process and its key elements is paramount to a product's success.

This section forms the core element of the IGCSE® Design & Technology course. You will look at the process that many product designers go through to produce items, as well as a number of issues that designers must consider when they are designing a product.

STARTING POINTS

- What is a design brief?
- What do you do with a design specification?
- Why is research an important aspect of designing?
- What types of media are used for presenting ideas to different clients?
- How do you take an idea and turn it into a product?
- How can IT be used to help the design process?
- What moral and ethical responsibilities do designers have?
- What is planned obselescence?

SECTION CONTENTS

1·1 Getting started
1·2 Design ideas and techniques
1·3 Making
1·4 Evaluation
1·5 Health and safety
1·6 Use of technology
1·7 Design & Technology in society
1·8 Product design application
1·9 Environment and sustainability

1 Product Design

1·1 Getting started

1·1·1 Observing a need

LEARNING OBJECTIVES

By the end of this section you should:
✓ understand a design need or situation
✓ understand the importance of having a client.

DESIGN NEED

Most aspects of design are pushed by commercial & economic desires by companies to make money. However, this is not always the case. Many designers and inventors are driven because they are dissatisfied by existing products and wish to improve them: they want to help other people, or they use new technology and science to create products that may make our lives easier.

Designers are individuals or groups that are creators of products. They may work for small companies or they may work as part of a large multinational organisation. Designers usually make their products for a **client**. A client is usually a company or an entrepreneur who may finance a new product.

The starting point for designers' work often comes from two areas: either a need/situation for a new product that will solve a particular problem, or to develop an existing design/product to improve on an old one. The finished product is often called the **outcome**. Designers also have to be clear about who they are designing for. A term used for this is the **end user**.

Designers work with their client to establish a **design brief**. This is usually the starting point of the **design process**.

In school, an example of a problem may be that your school librarian (client) requires a seat (outcome) for a particular area of your school library for students (user) to use. Seating is not new but this need is a specific seat for a specific place.

Often, the best products come from a genuine problem or need. For example, producing book stands for a new library in school or designing an electronic scoring device for team sports like basketball. This can be either a new solution or the development of an existing product for a particular client or user.

There are products, systems and environments all around us that could be improved or have particular issues that could be solved. Some themes may include:

- problems in your home
  - sorting out and organising a desk
- sport and recreation
  - equipment storage
- support for specific user groups
  - aids to support domestic tasks
- safety
  - signage and instructions
- pets and animals
  - creating a play gym for small pets

△ A common problem at home and work is a messy desk.

THE IMPORTANCE OF A CLIENT AND USER OR USER GROUP

When you have a theme, need, problem or situation you will need to analyse it in detail to understand it fully. If you do this properly you will have a real understanding of the issues and constraints involved.

It is best to have a client you are designing for who can engage in the process and give you genuine insight and **feedback** on your work and analysis. It is also good to have a **target market** or **user group** in mind. This will be the end user or main customer for the product you are designing. Take your time to discuss your thoughts with your teacher, client and the end user.

A quick brainstorm is a good way of putting down on paper your initial thoughts and findings:

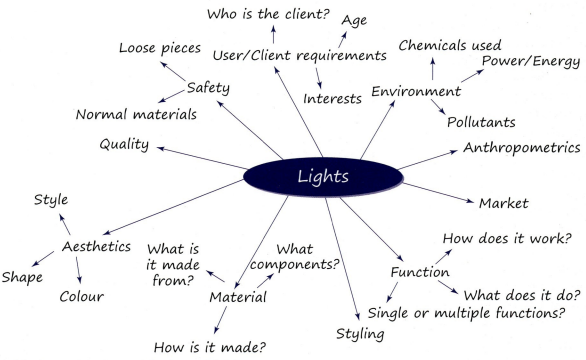

△ A brainstorm based around the theme of lighting

TOP TIP

When creating a brief and looking for opportunities, it helps to have a theme that you are genuinely interested in.

SKILLS ACTIVITY

Look at this photograph.

In the image, you may see many problems to solve.
Ask yourself:

- what is the problem?
- why do we have this problem?
- who does the problem involve?
- how could I make it better?

▷ This container is full of multiple leads from mobile devices.

> **KEY TERMS**
>
> **client:** the person or company you are working for to create a solution
>
> **design brief:** a statement of your design intent or a description of what the product will be like
>
> **design process:** a logical problem-solving process of designing a product
>
> **end user:** someone who uses or will use your product
>
> **feedback:** information and opinions about the work or product from people who have seen or used it
>
> **life cycle:** all the stages of a product's existence from its design and creation, through its use, to the end of its 'life'
>
> **outcome:** the final project or product
>
> **sustainability:** the ability of materials to be used wisely with little or no negative impact on the environment
>
> **target market:** the group of people you hope will use your product
>
> **user group:** the type of people who use or will use your product

1·1·2 Design brief and specification

LEARNING OBJECTIVES

By the end of this section you should:
✓ understand the importance of a design brief and what it should include
✓ be able to analyse the brief to fully understand what you need to do.

DESIGN BRIEF

Once you have a clear idea of the problem or situation you would like to solve it is time to write a design brief with your client. This is the first stage of the design process.

Working closely with the client is important in order to get a full understanding of what is required. The design brief is a statement that sets out what you would like to design and make.

The brief may include details such as:

- what the problem is
- where the problem is
- the function of the product
- who is going to use your product
- how often and when the product will be used
- what type of product you are going to design and make
- what materials will be used
- health and safety considerations.

You may add other key points driven by further analysis of the problem.

Here is another example of a design brief using a Chinese New Year theme:

Design brief
I am going to design and make a piece of modern jewellery for my cousin who will be celebrating Chinese New Year very soon. The jewellery will be worn when she meets family in the first days of this cultural celebration. Traditionally people wear red clothing during this time and she would like a piece of jewellery to complement her outfit.

Once your brief is established, it will need to be analysed so everyone has a full understanding of what will be designed. A very useful task is a 'P's analysis, or marketing mix. Consider the following:

- **people:** the users of the product and those involved with the project
- **places:** where the product is going to be used

- **properties**: style, finish and materials to be used
- **products**: availability of similar products on the market now
- **price**: selling price
- **process**: the process needed to produce and distribute the product
- **physical evidence**: ways in which the product can be demonstrated or shown to the target market before they purchase
- **promotion**: how the product will advertised to the target market.

SPECIFICATION

The **specification** is a document that develops the brief. It contains the key points that have been researched about the problem, and any observations from discussions with the client and user or user groups. In addition, any other initial **research** is included to help guide the specification. It is usually a very detailed document, and it is very important as it is the basis for the designs. Sometimes designers produce an initial specification, which lists key **constraints**, and then develop this into a full specification later. Once the specification is set, designers will check it at regular intervals to make sure they do not deviate from the original brief.

Specification

- Function – What the product does. It may have several functions.
  (Brief Analysis/Situation Analysis)
- Performance – How well it performs its functions.
  (Brief Analysis/Situation Analysis/Focus Group)
- Market – What makes the product appropriate for the target market?
  (Focus group/Questionnaires/Product Analysis)
- Aesthetics – What aesthetic qualities will the product need to have?
  (Situation Analysis/Focus Group/Questionnaire/Product Analysis/Aesthetic Influences)
- Quality – Define the quality levels that aspects of the product will need to achieve.
  (Situation Analysis/Focus Group/Product Analysis)
- Safety – What factors will make the product 'safe'?
  (Focus Group/Product Analysis/Safety Research)
- Environment – How can the product be produced with minimal effect on the environment?
  (Materials/Finishes/Manufacturing/Assembly)
- Material Properties – What properties do the materials used in the product need to have?
  (Situation Analysis/Focus Group/Product Analysis)
- Size – What are the key sizes important for this product, including anthropometric data? (Product Analysis/Anthropometric Data)
- Cost – How important is cost to the client/target market?
  (Situation Analysis/Focus Group/Questionnaires)
- Packaging – What must the packaging achieve?
  (Brief Analysis/Situation Analysis/Packaging Research)

A Specification:
- is a list of requirements
- sets out the key features of a product
- is a series of bullet points
- is clear and succinct
- is precise and specific using data wherever possible
- relates to the research
- is the criteria for evaluation
- is used throughout the design process.

△ An example of a specification.

TOP TIP

When creating a specification, a good starting point would be to use these acronyms: CAFEQUE or ACCESSFM. Both of these provide many of the points a designer must consider.

Cost	Aesthetics
Appearance	Customer
Function	Cost
Ergonomics	Ergonomics
Quality	Size
User	Safety
Environment	Function
	Manufacture

> **KEY TERMS**
>
> **constraint:** something that may limit aspects of your design
>
> **research:** the process of gathering new information about something
>
> **specification:** a list of key points describing the construction, materials and appearance etc. that your design must have

1·1·3 Research

> **LEARNING OBJECTIVES**
> By the end of this section you should:
> ✓ be able to put together a research plan
> ✓ be able to identify different research methods
> ✓ be able to identify constraints imposed by knowledge, resources and external sources
> ✓ be able to gather, order and assess information relevant to the solution
> ✓ be able to produce and interpret data (for example, diagrams, flow charts, graphs, test results).

THE IMPORTANCE OF RESEARCH IN DESIGN

An important part of the design process is the initial research a designer does to collect all the relevant information they need in order to create a new product. This information forms the foundation of the development of any new product and is a vital part of the design development system. Research can be time-consuming and often only a small portion of the research is actually used in the final design process, as much of the research is sometimes irrelevant or not appropriate. **Brainstorming** is a useful starting point and it is helpful to get some initial thoughts down on paper.

Research is often carried out by market research companies who invest a great deal of money and time in discovering exactly what people want. They do this in various ways, including interviewing people and using **surveys**. Surveys can be done over the internet, by telephone, in person or via post. The types of information collected can be **quantitative** and give data to analyse, or **qualitative** where opinions are sought. This research helps designers to develop new products and improve existing ones.

ORGANISING RESEARCH

Research needs to be gathered, ordered, analysed and presented. Make sure the research is useful to your theme/brief and specification and discard irrelevent or useless information.

All sections of the research must be evaluated and conclusions drawn from each different type of research (for example, observation, data collection, user interview, product analysis). At the end of each piece of research, create a box and write in your conclusions/summary (see example over the page).

There are two distinct types of research:

Primary research is where you carry out the research and collect the data and information for yourself. It may include:

- looking at existing or similar products and comparing them
- looking at how existing products are manufactured and assembled
- discussing the problem further with your client and/or user group

- broadly observing similar products being used by the general public
- observing the user group specifically with similar products
- looking at what materials are available
- finding out which processes you have access to in school
- looking at the sizes of people; for example, seating height
- measuring dimensions of fixtures and fittings
- measuring dimensions of items that are linked to your proposed product
- deciding on the **finishes** available to you.

Secondary research is where you gather information from other people. The source is not your own. For example:

- designers' work
- magazines
- data books
- internet
- **anthropometric** (people size) data.

It is also important to note that research is carried out throughout the process of designing as well as at the outset.

> **TOP TIP**
>
> It is very important to plan. Clearly define which research tasks need to be completed and start working through each task. An example is given below:

Research tasks	Why?	How?
Find out how much the client will pay for the lamp.	To be aware of potential budget.	Interview and questionnaire
Find out where the lamp will be used.	To design a lamp to meet a room theme or style.	Interview and questionnaire
Find out what the different packaging methods are.	To find out what is the most suitable style of packaging for the product.	Internet and books
Find out what the different types of light fittings are.	To fully understand the types of light fitting available and most suitable for the task.	Internet and books
Find out which materials are most environmentally friendly for the product.	To try and make the product sustainable.	Internet and books
Find out what safety issues need to be considered.	Customer safety and safety in manufacture.	Internet and books

A research plan is a good method of clarifying the task you need to complete. For example, what do you need to do, how you are going to do it, and where do you need to get the necessary information from?

EXISTING PRODUCTS AND PRODUCT ANALYSIS

An important stage in the design process is research into existing products. Designers need to investigate and analyse products and ask the question, 'Can I improve the design in any way? If so, how?' When looking at an existing product, the following needs to be taken into consideration:

Materials
What materials is it made from?
What properties have these materials got?

Aesthetics
Is the product designed in a specific style?
Does it look attractive?
What are your feelings about this product?

Target market
Who is it for?
What age group?
What makes it appeal to a target market?

Manufacture
How has this product been made?
What methods of construction were used?
Is this batch a one-off or mass produced?

Environmental concerns
Does this product harm the environment in any way?
Is it made from any recycled materials?

Cost
What is the making price of this object (if known)?
What is the selling price?
Does the quality have an influence on the price?

Ergonomics
Has the product been made to suit the user?
Is it easy to use?
Is it easy to adjust?
Is it comfortable to use?

Function
What does this product do?
How does it work?
Does it work well?
Does it have special features?

△ Example questions to consider when analysing existing products.

Looking at the work of other designers, both past and present, can be very good inspiration. However, care must be taken not to copy or plagiarise other people's work.

TOP TIP

When carrying out primary and secondary research you often realise that there can be constraints or limitations on your designs. For example, if you are looking at lighting design you may only have access to a specific bulb holder, which would limit your options. It is important to consider all your constraints and limitations, so make a list of these before you start designing.

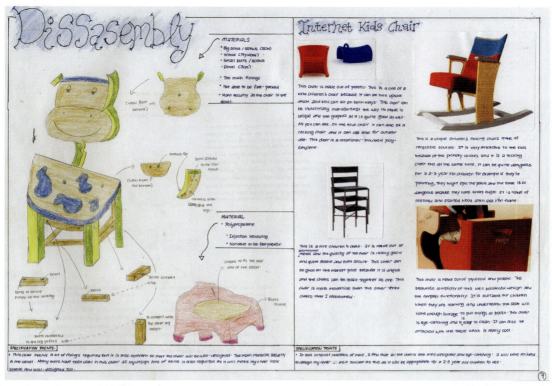

△ Above is an example of a student's portfolio sheet showing the disassembly of an existing product and research into other products of that type. (Please note that the student has spelt 'disassembly' incorrectly. You must ensure that you spell words correctly in your work.)

KNOWLEDGE CHECK

Young people often carry many items in their bags when they go to school and walk to their classrooms.
A designer has been asked to design an alternative carrying system for young people.
1. List four functions the designer must consider for this product. (4)
2. List six types of research the designer could carry out to fully understand the problem. (6)

KEY TERMS

anthropometrics data: information about the measurements and properties of the human body (for example hand sizes of 15-year-olds)

brainstorming: the process of pulling apart an idea, theme or problem in order to analyse it, discuss it and to generate new ideas

finish: a substance that is used to produce the protective layer or texture colour for the surface of a product to make it last and look good

primary research: new research that you carried out yourself

qualitative data: information in the form of opinions

quantitative data: information that can be easily expressed and analysed in a numerical form

secondary research: research based on other people's work

surveys: a set of questions asked to the public or to user groups in order to gather specific information to help develop products and services

1·1·4 Initiating and developing ideas, and recording data

LEARNING OBJECTIVES
By the end of this unit you should:
✓ be able to extract relevant information from sources, interpret and record information and data.

DATA ANALYSIS AND INTERPRETATION

It is extremely important to be able to take all your research findings and extract the most relevant information that relates to your design brief and specification. When gathering research it is common to have piles of information and data that may not fully relate to your problem or opportunity. This does not mean it is not important; it simply may not be relevant at this stage. For example, you may have gathered information about a range of products that are already for sale on the market. The products may be made from a wide range of materials but you are going to use a specific material. It does not mean the other products are not relevant, as details such as styling and fixings may be useful.

The data and information you are interpreting may be in the form of:

- questionnaires to gather information about potential users of the product you will design
- interviews with users of a product
- observations of people using a product
- experiments and testing of materials, finishes and working products
- disassembly of existing products to see how they are made
- testing products for durability, function and ease of use.

There may be many more forms of information that you have gathered, but the key is to take what is useful and relevant to your project and problem, and leave what is not. This may seem like a waste of valuable time but, by putting some information aside, you do have to think about what is not relevant to the design or problem. This is part of the design process and will ensure you have a full understanding of the problem and situation.

Once you have a full understanding and all the information available, you are in a position to start the fun activity of creating solutions. There needs to be evidence of primary and secondary research. However, primary is often more useful.

SKILLS ACTIVITY

This diagram demonstrates the possible range of resources available to the designer. Can you think of any more to add to the diagram?

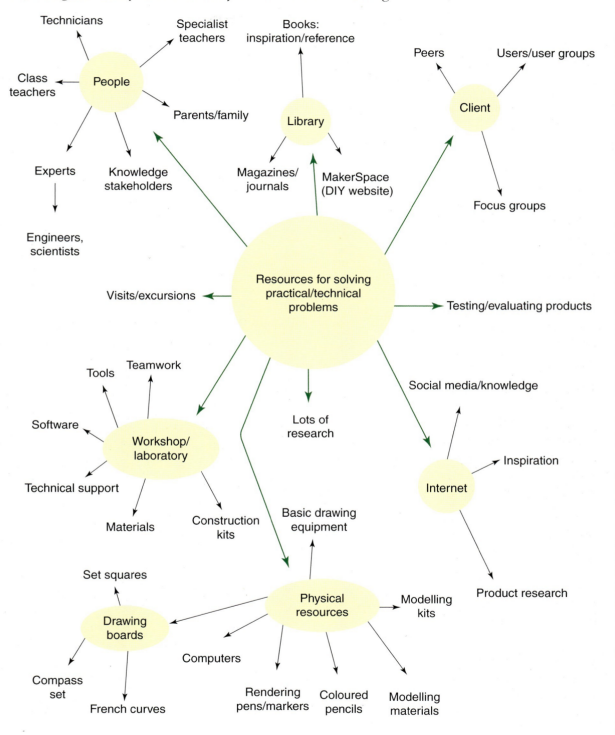

1·2 Design ideas and techniques

1·2·1 Generating possible ideas

> **LEARNING OBJECTIVES**
> By the end of this unit you should:
> ✓ be able to generate and record a range of design ideas
> ✓ be able to identify what resources are needed to solve problems.

GENERATING DESIGN IDEAS

At this stage of the process designers will have a very clear idea of the design problem. They will have discussed the problem with a client and potential users, undertaken some research into the problem, interpreted any data that has been gathered, and then written a design specification for a possible solution.

Next, designers will start generating a wide range of imaginative preliminary ideas for what the solution may look like and, if it is a system (for example, electronic or mechanical), what it is supposed to do. This is an extremely important and fundamental part of designing as it allows the designer to demonstrate creativity in solving problems. It is also a lot of fun!

Ideas are often freehand and manually drawn but can also be digitally produced. These initial ideas are usually quick outline ideas and a good way of communicating thoughts and ideas to the client and user.

It is critical that all your ideas are recorded so they can be referred to later, which will allow you to see how your ideas have progressed. You may, for example, produce electronic circuits on a piece of software, draw ideas on a sketchbook, or use a tablet to put together some ideas.

Strategies for generating ideas

Here are some different strategies for starting your ideas, explained in detail below:

- moodboards
- SCAMPER
- random shapes
- freehand sketches
- modelling.

△ Sketchbooks can be used for recording ideas and developing designs.

Moodboards

A good starting point is to generate a mood or image board. This is a page of images relating to your product, its theme, intended users, and so on. You may have a theme, for example children's teaching aids, timing devices, sports merchandise. All three themes would have a very different range of colours, users and locations. The collection of images will help stimulate ideas in terms of colour schemes and styling.

In addition, to help you start generating ideas, you may want to think of generic themes such as nature and natural surroundings, geometric shapes, work of past designers or design/art movements. This technique can get you started when you may have limited ideas initially. However, always relate this thinking or designing strategy to your problem and client requirements.

△ A designer's moodboard for an ideas generation task, showing childlike themes. It was used to generate ideas for a storage box for children's toys.

Scamper

This tool helps generate ideas by encouraging a focus on improving existing products:

S – Substitute

C – Combine

A – Adapt

M – Modify

P – Put to another use

E – Eliminate

R – Reverse

Applying these actions to an existing product or idea can be very useful when developing or exploring an idea. See the SCAMPER diagram in unit 1·8·3 for further information.

SKILLS ACTIVITY

Look at a product that is near to you right now. Sketch it on a piece of paper. Choose one of the SCAMPER strategies above to develop the product. Repeat the exercise for a range of SCAMPER strategies and see how your initial product could be developed.

Random shapes

When generating ideas, designers initially start with the overall shape or form of the item they are designing. They then start looking at the smaller aspects of the design.

Ideas can be sketches, detailed views or **renderings**. Use the 'Let your pencil do the walking' technique for shapes. Draw random lines on a sheet of paper. Pick out interesting shapes that could form the basis of a design.

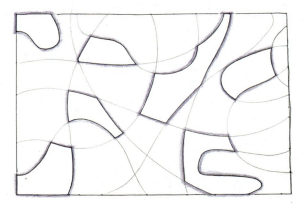

◁ An example of rough lines that are highlighted to create interesting shapes. They can be used to form the basis of design.

Freehand sketches

Initial freehand sketches are a great way of generating and recording initial design ideas.

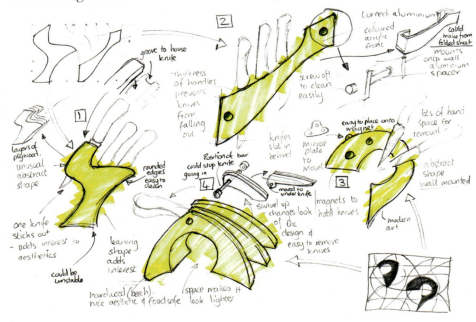

△ Here, a student has used freehand sketches to put their design ideas onto paper and show the development of their ideas.

You can use a wide range of media and techniques to present these ideas. Media can include:
- pencils
- markers
- ballpoint pens
- felt tip pens
- crayons
- pastels
- chalks
- coloured/textured paper.

This is a very creative and innovative part of the process and you should try a range of techniques. There are many different techniques for illustrating your ideas:

- **2D and 3D sketches**
  These are quick and are a good way to explore ideas. It is a good idea to use annotation to support your ideas and help explain your thinking. The annotation may include ideas for materials, manufacturing methods, finish of materials using colour and texture, ideas, and any thoughts relating to the brief, specification and user.
- **Rendering**
  Ideas that are rendered (colour and texture added to reflect the final look of the product) are usually 3D and are more realistic. They are completed by hand, or on a PC or tablet computer. There are many pieces of software and applications for doing this. Some popular ones in schools are AutoDesk® Fusion 360™, PTC Creo, formerly known as Pro/ENGINEER, and Sketch Up. This is known as computer-aided design (CAD).

△ A 3D outline image of a lamp

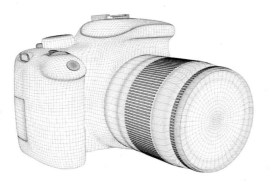
△ A camera rendered in 3D using CAD software

Drawing techniques such as **orthographic projection** or working drawings are used to present ideas formally. They are usually to **scale**, have **dimensions** or sizes on them, and include details of materials and finishes.

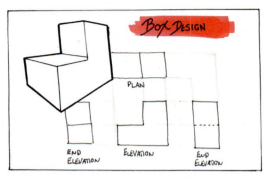 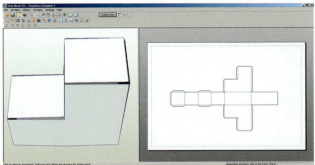

△ 3D design ideas turned into CAD drawings

Modelling

Modelling is another way of creating ideas. The manipulation of 'soft' materials into working models can be beneficial to see if the concept works. If developing an electronic or mechanical system such as an electronic scoring device, a systems approach to the ideas may be required to work out the product's operation, as well as designing the overall casing to house the parts. See unit 1·6·2 for more information on this.

This system will probably need to be designed on a piece of software, but it could also be created by using existing circuit designs or mechanical systems and combining them to produce a solution that meets the brief.

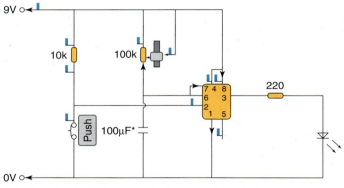

△ A circuit made from electronic circuit modelling software

△ A systems diagram of an electronic dispensing machine

TECHNICAL VOCABULARY

The use of words and terminology is very important when trying to develop ideas and annotating them. Designers need to take ideas and discuss them with clients and user groups and the descriptions need to be correct.

TOP TIP
When you are sketching, modelling or rendering to generate ideas it is important to constantly evaluate them against your design brief and specification, so that you can ensure you are designing what you are supposed to be designing.

CONSIDERING RESOURCES NEEDED TO HELP SOLVE PROBLEMS
When designing and solving practical and technical problems, it's important to use all available resources. Designers need to have a full understanding of the problem by gaining information from many sources. Refer to the mind map in unit 1·14 which provides examples of resources that may help you when designing.

SKILLS ACTIVITY
On an A4 piece of paper, produce a set of freehand sketches and annotate them for the following problem:

Your headteacher wishes to redesign the logos of the school house or team system. Come up with a range of ideas for merchandise and advertising. (Keep your work as you will need it again in unit 1·2·4.)

KNOWLEDGE CHECK
1. Name three techniques to help you generate ideas. (3)
2. Explain why it is important to record and keep the ideas you generate. (5)

KEY TERMS

dimensions: the measurements of the size of an object, usually shown in mm

orthographic projection: a two-dimensional drawing that represents a three-dimensional object, showing several elevations and dimensions

rendering: adding line tone, colour and texture to a drawing to make it look realistic

scale: the ratio between the size of something real and a model or drawing of it, that has either been enlarged or reduced

1·2·2 Use of media for mock-ups

LEARNING OBJECTIVES
By the end of this unit you should:
✓ be able to use a variety of media and equipment to produce models and mock-ups as a means of testing a solution.

THE IMPORTANCE OF USING MEDIA FOR MOCK-UPS

A very important part of the design process and development of ideas is creating whole and/or part models of their proposed solutions. **Scale** or **life-size** models can be made into 2D shapes or 3D forms. It is sometimes far easier to visualise and get a better understanding of a design with a model than with a drawing. Models allow testing of mechanical functions and evaluation of their form. They also allow you to think through construction techniques, size, proportion and materials as well as any assembly details, materials, finishes and additional components that may be added to the design, such as switches and bulbs.

MATERIALS AND EQUIPMENT

A wide variety of media can be used in constructing mock-ups. Initial quick **sketch models**, made of simple soft materials, are a good starting point to see the shape or form of a design. Materials such as card and paper can all be easily cut and manipulated. More detailed models, made from harder, more accurate materials such as Medium Density Fibreboard (MDF), can be made at a later stage to present to clients and team members and to also fully test concepts. Models may take the form of software or kit modelling for different projects, such as software modelling in electronics and mechanisms modelling with LEGO®.

Some materials for modelling include:
- paper
- card
- toothpicks
- corrugated card
- chopsticks
- wire
- drinking straws
- balsa wood
- clay
- **polymorph**
- foam board
- styrofoam
- string.

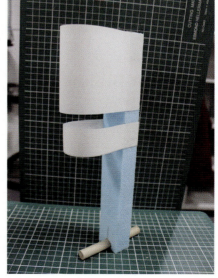

△ Styrofoam models are cheap to make.

Models can be made with basic tools. More sophisticated models using hard materials can be made using workshop equipment and computer-aided manufacture (CAM).

Useful equipment includes:
- double-sided tape
- scissors
- craft knife
- cutting mat
- safety rule
- glue gun/sticks
- masking tape
- circle cutter.

TOP TIP

Add modelling examples and their development to your portfolio and annotate them to help you explain the progress of your ideas, and also to evaluate them. Include written evaluations and attribute-analysis diagrams. This type of analysis looks at a number of points of the design and grades them on a scale. The analysis that is nearest to a circle indicates that the design has the best attributes against the given criteria.

MODEL 1

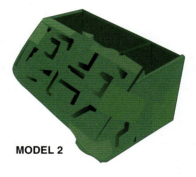

MODEL 2

This model is not very interesting. It has a simple and basic shape. There is no opening for the wire of the lamp to come out.

The design of this model is not very engaging. To improve it, the front design will be re-worked to make it more modern and appealing. There are openings at the side of the design, to allow the wire of the lamp to come through.

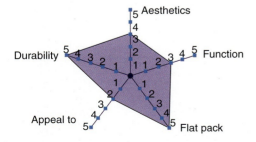

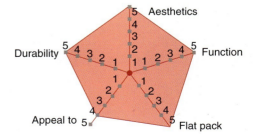

△ Using attribute analysis and star diagrams, such as above, is a useful way of evaluating a product's success.

COMPUTER-AIDED DESIGN (CAD) MODELS

Computer models are a very quick and easy way of developing a three-dimensional form of a design. Software allows designers to create models quickly and also gives them the opportunity to change, refine and develop concept ideas easily. Sophisticated software allows designs to be rendered, shaded, textures added, modified and viewed from all angles. With rapid prototyping/3D printing, CAD models can be printed relatively quickly to produce high quality models and working components using computer-aided manufacture (CAM). This will be covered in more detail in unit 1·6.

△ CAD drawings can show a range of edges

CONSTRUCTION KITS

Construction kits such as LEGO®, which have a range of parts to assemble into various designs, are extremely valuable for modelling. Kits can be used as a method of quickly creating models and also to test mechanical features of a design. Electronic kits can also be an easy option for testing any additional detailing in a design, such as basic wiring.

△ Electronic kit

△ LEGO® kit

TOP TIP

Search for videos on the internet with keywords such as 'product design modelling' and view the range of ways simple and complex models are made.

TOP TIP

Modelling is an extremely important aspect of developing designs. Do not be afraid to make many models to ensure you have thought through your design and how it is going to be made. If possible make full-scale models to really appreciate what you are designing.

Make quick models for concept ideas and development. Make more detailed, permanent models for sharing your ideas with clients and peers for evaluation and critique. Remember these models are prototypes and not the final product.

SKILLS ACTIVITY

Design a chair. Make a simple model from a single A5 sheet of paper.

KNOWLEDGE CHECK

1. Name a material that is commonly used in car design to model the initial shape of the vehicle. State why you think the material is used. (2)
2. A school stationery shop wishes to sell a range of common stationery as a promotional set for your school. Outline a method of producing a prototype of your solution. (4)

KEY TERMS

life-size model: a model that is the same size of the intended outcome

polymorph: a smart material that can be heated in water to 62 degrees and shaped. Once cooled it can be worked like a traditional material, with the advantage that it can be reheated and reshaped.

scale model: a model that is proportional in size to the final outcome

sketch model *or* **mock-up:** a quick, simple model of a design

1·2·3 Generation of possible ideas

LEARNING OBJECTIVES
By the end of this unit you should:
✓ be able to recognise the need for continual appraisal and evaluation
✓ be able to relate appraisal and evaluation to the specification.

THE IMPORTANCE OF CONTINUAL APPRAISAL

Evaluating and appraising initial design ideas and, indeed, further ideas is a continual task. Designers will always look to their specification, design brief, client requirements, research and the end users to check if the ideas being produced meet all the requirements set out. This can be a challenging but interesting task. Meeting all these criteria can take time and require many developments of a design idea. There are a number of design evaluation methodologies you can use.

Pass-fail analysis

Pass-fail analysis entails creating bullet-point lists of all the design's criteria and then making a pass or fail judgement for each one. Explaining your reasons for each judgement is also helpful as it makes you reflect on your decisions. Here is an example:

Specification point	Achieved: Yes or No?	How have I achieved the specification point?
The 3D display I am producing must be able to be packed flat.	Yes	Using slots in the material for self-assembly and dismantling.

SWOT analysis
- **S**trengths
- **W**eaknesses
- **O**pportunities
- **T**hreats

Score card

Each specification point is given a score from 1 to 5, with 1 being the highest and 5 being the lowest. Here is an example from a board game:

Point	Score	Outcome/Action
The instructions for the board game are clear for the user.	1 2 3 **4** 5 Clear Unclear	The instructions need rewording and simplifying.

Traffic lights

This is a visual way of understanding and reviewing progress of each part of the design criteria. With this strategy the criteria are coloured red, amber/yellow or green.

- Green = go
- Amber/Yellow = working but needs developing
- Red = stop

Star diagrams with specific criteria

Star diagrams are useful when you have a range of criteria and wish to judge and then give an overall evaluation of the product. Each criteria is scaled with the larger number on the outside indicating a positive result. All the points are joined. A good result would have all the lines on the outer edge of the star. Any lines not on the outer edge need to be addressed in the design.

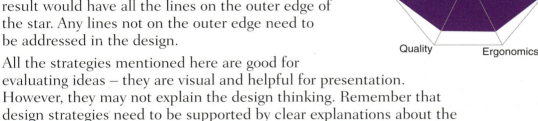

All the strategies mentioned here are good for evaluating ideas – they are visual and helpful for presentation. However, they may not explain the design thinking. Remember that design strategies need to be supported by clear explanations about the judgements being made.

TOP TIP

Designers need to be very good at evaluation. In industry, designers are developing products for markets, not for themselves! This often means designers could be creating products that they do not like.

When taking part in design activities as part of your studies, it is important to choose a theme and topic that you are interested in and will be fully engaged in over the duration of the project. If the theme is given to you, develop it to suit your interests.

SKILLS ACTIVITY

Look closely at a product you are familiar with. Explain how you would test and evaluate the product.

Why should evaluation take place at all stages of the designing process?

1·2·4 Communicating design ideas

LEARNING OBJECTIVES
By the end of this unit you should:
✓ be aware of a range of methods for communicating design ideas.

METHODS OF COMMUNICATING IDEAS

At this stage of the process designers usually have an idea of what they are going to make. The way they communicate these ideas to the client and users is important, so they have a clear understanding of the designer's thinking and solutions. When communicating, it helps to think about these questions:

- Why is the communication needed?
- Who is the communication aimed at?
- What method/type of communication would be most effective?
- What additional information needs to be communicated?
- What changes or decisions need to be communicated?
- What positive or negative points need to be communicated?
- What reasons for decisions and changes need to be communicated?

Communication can take many forms. The one you choose will depend on what you are trying to communicate, but some useful forms are:

- **Rough sketches**
    - models made from a variety of materials
    - written explanations and annotations to support sketches
    - mathematical representations (graphs, charts).

△ A freehand pencil sketch

For example, if you are discussing possible solutions with a user at an early stage, rough sketches may be most suitable with annotation to explain your design thinking. If you wish to explain to the client how a mechanical function may work in a design, a model may be more suitable. If you wish to convey the materials and finish of a design then a rendered and shaded sketch or CAD drawing may be most suitable.

When ideas start to take shape, more formal techniques can be used to make the drawings clearer and more realistic.

- **Isometric drawings**
    These are 3D images that look realistic, such as the toaster below:

- **Perspective drawings**
  These are 3D images where the horizon can be changed for different views. Below is a perspective drawing of a torch:

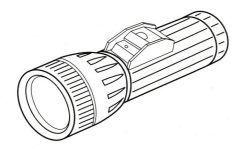

- **Exploded views**
  Exploded views are drawings that show all of the separate parts of an assembly and how they would link together, such as the computer mouse below:

All of these methods should usually show the materials to be used. Various media can be used for this: crayons, markers, pastels, coloured pencils, watercolours, and so on. Often CAD drawings and computer software are used.

When communicating with clients or peers, the use of written explanations and notes can also be very important in supporting your ideas. Using technical vocabulary is very important for clarity and precision.

Working drawings – 2D drawings

When using drawings for manufacture a designer will use working drawings. This could include a drawing that shows the completed product and also sub-assembly drawings. Working drawings all have accurate dimensions, materials and tolerances to enable the manufacture of the product.

 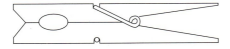

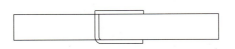

◁ An orthographic drawing of a clothes peg

PRESENTATION BOARD

A presentation board is a great way of putting together a combination of ideas and solutions on one page. It allows the client to see the design in different formats to aid understanding. The board may include models, hand-drawn ideas, computer-generated ideas, and written explanations. It is simply a collection of design work that has been created and put in one place.

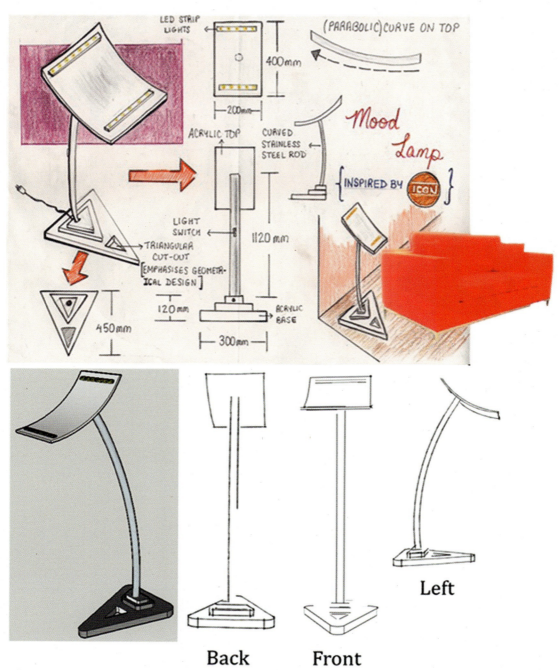

△ A presentation board for a lamp

SKILLS ACTIVITY

Using your sketches from the second Skills Activity in unit 1·2·1, develop your ideas using a broader range of presentation and modelling techniques to produce a presentation board that could be used to present to a client. A reminder of the problem is below:

Your headteacher wishes to redesign the logos for school sports teams. Come up with a range of ideas for merchandise and advertising.

The template opposite can be used as a guide for your presentation board.

A3 portfolio sample layout
This is not fixed and can be created by you as a designer!

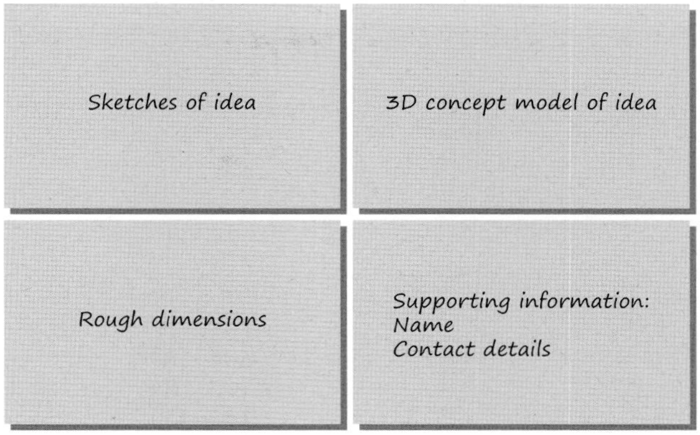

Sketches of idea	3D concept model of idea
Rough dimensions	Supporting information: Name Contact details

1·3 Making

1·3·1 Selection and organisation

LEARNING OBJECTIVES

By the end of this unit you should:
- ✓ be able to select and develop a solution based on time, cost, skill and resources
- ✓ understand that making needs to be planned and organised in detail.

CHOOSING THE FINAL SOLUTION

Once the final proposal has been evaluated against the specification, meets the clients' and users' needs, and has been developed, it is time to think about manufacture.

Design ideas may have been modelled at this stage using CAD software and by making sketch models out of some sort of material to get a scaled version of the design. 3D prototyping may have also been attempted (CAM). However, experimentation with the final materials that the product is to be made from may not yet have been attempted.

MATERIALS SELECTION

During the development stage, final materials are often chosen for their properties: finish (aesthetics), **costs**, availability, and whether machines and equipment are available to make the pieces required in the design. Other components such as **fixtures and fittings** may also need to be sourced and costed to ensure the product meets its target price. In addition, having the **skills** to manufacture the specific parts is essential.

Testing and combining the materials chosen with all the extra components, in terms of joining methods, finishes and shaping methods, is a great way of determining if the material choices, processes and skills are the right ones. If the material and component, combinations do not fit well together, the product can be developed further to ensure the components are suitable for the job they are intended.

MATERIALS TESTING

At this stage, designers may test materials to establish if they are fit for purpose, that is, that they do the job they are supposed to do. There is a range of scientific testing, including:

- density test
- heat/melting point test
- tensile test
- izod impact test
- conductivity/insulation test
- hardness test.

SKILLS ACTIVITY

Research these material testing terms and explain how the individual tests may be useful for materials selection in your design work.

It is possible to carry out simplified materials tests in a workshop with some basic tools and test rigs.

In addition to hard materials testing, designers may check the suitability of electronic components, mechanical components and graphics-based materials. In an industrial context this may be completed through software. In school, you can also do this with appropriate construction kits.

△ Tensile testing taking place in a commercial testing rig to measure the amount of stretching the material can take before breaking.

MATERIALS AND CUTTING LIST

A major part of planning the making and manufacturing activity is to list all materials and components that make up the design. This is a useful document to work from.

Materials list

Name	Material	Quantity	Length (mm)	Width (mm)	Thickness (mm)
Front-base	Acrylic	1	275	200	3
Primary function separator	Acrylic	1	275	100	3
Secondary function separator	Acrylic	1	104	100	3
Back	Acrylic	1	275	103	3
Cover	Acrylic	1	222	85	3
Side	Meranti	1	95	100	10
Side template	Acrylic	1	95	100	3

△ An example materials list

PLANNING FOR MANUFACTURE

Once materials, processes and additional components have been selected and proven to work together, the designer will plan for manufacture. To do this efficiently, designers will carefully consider and plan each step of the process to ensure everything is completed in the correct order. There are a number of ways to do this and the following examples are given for consideration.

Flowcharts

A **flowchart** works logically through a sequence of operations, such as each step of the making activity. Each instruction is in a rectangular box and decisions (which can be quality and safety checks) are in diamond-shaped boxes. The start and end of a sequence is in a terminal (start and stop) point.

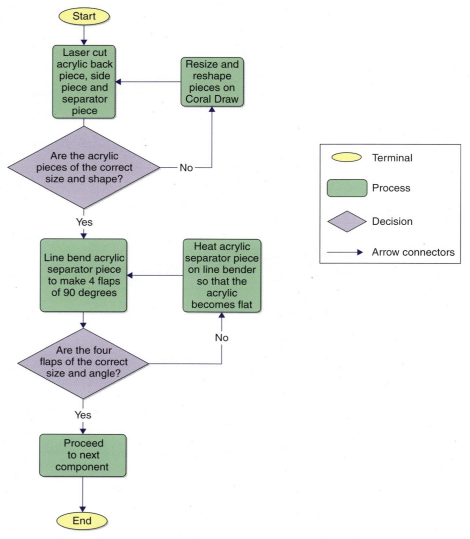

△ An example of a flow chart

Manufacturing plan

A manufacturing plan can be more detailed than a flowchart. It is usually presented in a table. Instructions for each stage of manufacture are listed such as marking out, cutting, wasting, finishing, fitting, assembly and testing. In addition, associated quality control checks, time allocation, health and safety considerations, and tooling/machinery are listed to make a comprehensive plan. The following is an example of a manufacturing plan and possible headings that you might include.

Process	Tools/ Machines needed	Quality	Quality check	Health and safety (risk assessment)	Time	Comments
1. Check that the materials used for the lamp production have the amount needed – laser ply, acrylic rod, threaded bar, nuts, polypropylene. Check against the materials list.	Ruler	Inspect the materials to ensure there is no damage that would decrease the aesthetics and function. Double check that there are enough materials.	All the materials should be of consistent quality. Check for blemishes, scratches, chips and marks.	Gloves for sharp edges Clean area to avoid any damage to materials or trip hazards Good storage	15 minutes	Ensure all materials are available before manufacturing starts.

SEQUENTIAL INSTRUCTIONS

Sequential instructions are clear drawings (usually with minimal text) that clearly demonstrate the steps in making something or putting something together. Many companies use them for products purchased in flatpack or kit form to be assembled at home. Companies such as LEGO®, Airfix and IKEA produce detailed and clear guides to aid assembly. The advantage of this type of instruction is that they are multinational as there is no use of language.

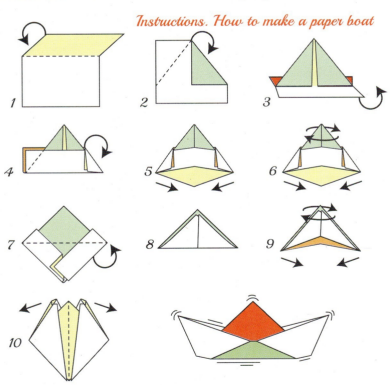

△ An example of sequential instructions for making a paper boat. Each step is clearly numbered.

TOP TIP

This stage of making is critical: the person making the product needs to have a clear understanding of all materials, equipment, processes, testing and checking that should take place to make the product. Take time to give careful consideration to this stage.

SKILLS ACTIVITY

Brushing your teeth is a function you carry out every morning and evening. Produce a flow chart that plans each step of this simple activity.

KNOWLEDGE CHECK

1. Name five things to consider when making your material selection. (5)
2. List five types of fixtures and fittings you may include in a design. (5)

KEY TERMS

costs: the amount of money needed to manufacture a product, including materials, tooling and labour

facilities: the equipment, processes and machines available to be used in the design or manufacture of a product

fixtures and fittings: standard components used to assemble a product that have been purchased rather than specially manufactured

flowchart: a diagram showing a sequence of operations in a process or work flow

skills: the abilities of the workforce or designer that need special training or expertise

1·3·2 Implementation and realisation

LEARNING OBJECTIVES
By the end of this unit you should:
- ✓ be able to demonstrate the correct procedures in preparation for making
- ✓ be able to draw accurately, mark out and test
- ✓ be able to select appropriate methods for manipulating a variety of materials.

PROCEDURES AND MAKING PREPARATION

At this stage of the designing process a solution to the problem has been found and the client and users are happy with the design. The next stage is to prepare for production. The drawings and research that have already taken place will be the foundation for this stage.

MARKING OUT

Marking out is where the raw materials are drawn on to mark out the sizes from the drawings in preparation for cutting.

When a piece of material has been prepared from a cutting list, it is always advisable to check that the sizes are correct (quality control). The material is then ready to mark out in preparation for machining, shaping, cutting, and so on. A **datum** needs to be prepared. This is an accurate line where all measurements will be taken from.

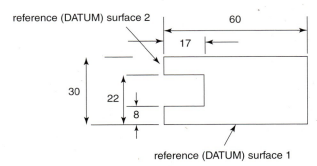

△ An example of datum faces on an orthographic drawing

Tools for marking out

On the next page is a table of common marking out tools in a basic workshop situation. When marking out, it is important to take your time, do it very accurately and check against your orthographic drawings. Making mistakes here could mean materials are wasted, which can cost valuable time and money. Always ensure:

- that sharp marking out tools are available
- that marking out guides are placed accurately on materials
- that you check for accuracy against the drawings.

Tool	Paper/card	Wood	Metal	Plastic
Pencil	✓	✓	✗	✗
Marking knife	✓	✓	✗	✓
Scriber	✗	✗	✓	✓
Marking gauge	✗	✓	✗	✗
Odd leg calipers	✗	✗	✓	✗
Crayon	✓	✗	✗	✓
Tri-square	✓	✓	✗	✓
Engineer's square	✗	✗	✓	✓
Compass	✓	✓	✗	✓
Templates	✓	✓	✓	✓
Engineer's ruler	✓	✓	✓	✓
Sliding bevel	✓	✓	✓	✓

TOP TIP

When marking out, follow the saying: 'check twice, cut once'.

Mark out your work, double check against the drawing – perhaps even have someone else check as well – then cut.

SELECTING APPROPRIATE PROCESSES FOR MAKING

The products all around us are made from a huge variety of materials. These materials are cut, shaped, formed, joined, fitted together, finished and tested, once again, by a large variety of machines and equipment.

Many of the processes available in industry will not be available in your school. However, having an awareness of industrial practices is important so that you appreciate commercial techniques. If a product is successful and goes into production, there will be many copies made, and knowing the machines that can do this quickly and cheaply is important.

Here is a table that loosely categorises materials with industrial processes. You will find more information in the relevant sections on Graphic Products, Systems and Control and Resistant Materials.

		Paper and card	Electronics	Wood	Plastics	Metals
Shaping	Laser cutting	✓	✗	✓	✓	✓
Cutting	Planing	✗	✗	✓	✗	✗
	Laser cutting	✓	✓	✓	✓	✓
	Shearing	✓	✓	✗	✗	✓
	Die cutting	✓	✗	✗	✗	✗
	Sawing	✗	✗	✓	✓	✓
Wasting	Abrading	✗	✗	✓	✓	✓
	Turning	✗	✗	✓	✓	✓
	Milling	✗	✗	✓	✓	✓
	Routing	✗	✗	✓	✓	✓
	Drilling	✗	✗	✓	✓	✓
Finishing	Laminating	✓	✓	✓	✓	✓
	Painting: – oil based	✓	✗	✓	✗	✓
	– solvent based	✗	✗	✓	✗	✓
	– water	✗	✗	✓	✗	✓
	Polishing	✗	✗	✓	✓	✓
	Powder coating	✗	✗	✗	✗	✓
Forming	Vacuum forming	✗	✗	✗	✓	✗
	Line bending	✗	✗	✗	✓	✗
Moulding	Injection moulding	✗	✗	✗	✓	✗
	Blow moulding	✗	✗	✗	✓	✗
	Rotational moulding	✗	✗	✗	✓	✗
	Extrusion	✗	✗	✗	✓	✓
	Compression moulding	✗	✗	✗	✓	✗

		Paper and card	Electronics	Wood	Plastics	Metals
Joining	Adhesives: – solvent	✗	✗	✗	✓	✗
	– superglue	✓	✓	✓	✓	✓
	– PVA	✗	✗	✓	✗	✗
	– hot melt glue	✓	✓	✗	✗	✗
	– epoxy	✓	✓	✓	✓	✓
	– contact	✓	✗	✗	✓	✗
	Welding	✗	✗	✗	✗	✓
	Ultrasonic welding	✗	✗	✗	✓	✗
	MIG welding	✗	✗	✗	✗	✓
Fitting	Rivets	✓	✓	✗	✓	✓
	Nuts and bolts	✓	✓	✓	✓	✓
	Screws	✗	✗	✓	✓	✓
	Nails	✗	✗	✓	✗	✗
	Knock down fittings (KD)	✗	✗	✓	✗	✗
	Staples	✓	✗	✓	✗	✗
Casting	Sand casting	✗	✗	✗	✗	✓
	Pressure die casting	✗	✗	✗	✗	✓
	Lost wax casting	✗	✗	✗	✗	✓
Printing	Gravure	✓	✓	✓	✓	✓
	Sublimation	✓	✓	✓	✓	✓
	Lithographic	✓	✗	✗	✗	✗
	Laser	✓	✗	✗	✗	✗
	Screen	✓	✗	✗	✓	✗
	Etching	✓	✓	✓	✓	✓

FIXING AND JOINING TECHNIQUES

When making a product, the different sub-assemblies (parts) need to be joined together. There are two main ways of joining parts: permanently or non-permanently. The type of fixing/joining methods used will depend on many variables, such as:

- Are the materials the same or different?
- Does the product need to be taken apart for maintenance or recycling?
- How strong does the join need to be?
- Should the join be visible or hidden?

Depending on the above, a decision will need to be made about the joining and fixing method used. The materials sections later will give more details. For example, you wish to join a piece of acrylic to some sheet metal. Will it need to be permanent? If so you may choose an adhesive such as epoxy resin. If non-permanent, then you may possibly use a nut and bolt made from steel or nylon. There will be many options, but you will need to make your decision based on specific criteria and which resources are available.

Standard components

Standard components are common components used in manufacturing and consumer items. For example, a battery operated torch comes with standard batteries. The manufacturer of the torch does not make the batteries. They design the product with the standard battery in mind. Another example of standard components used in manufacture are nuts and bolts. These are made by specialist companies who make millions of units, which keeps the cost low. Manufacturers purchase these components to assemble their products and save time and money. In the electronics industry, most components are standard and the combination of components create a range of circuits to perform different tasks.

△ Nuts and bolts are standard components used in manufacture.

Knock-down fittings

Knock-down fittings are standard components that are used to temporarily or permanently fix materials together. They are common in modern flat-pack furniture that consumers can build at home. Both manufacturers and consumers benefit from knock-down fittings as costs are reduced by putting the assembly element of the product construction in the hands of the consumer, therefore reducing manufacturing costs and the costs of the product.

△ An example of a knock-down fitting

SKILLS ACTIVITY

There are many types of knock-down fittings. Carry out research and identify as many knock-down fittings as you can find. You could use the internet at home or at school, a library, or resources in the classroom. List each example you find and state where they are commonly used. For example:

Knock-down fitting: plastic corner block
Typical use: kitchen cabinet carcasses

DESIGN IN ACTION

IKEA make considerable use of knock-down fittings in their furniture. Many of their products, including wardrobes, shelving units and bed frames, are flat-packed, meaning that the consumer can transport the furniture home in their own cars and not incur expensive delivery charges.

IKEA also use laminated manufactured boards, such as MDF and chipboard, in their furniture. This also reduces the costs for the consumer.

Building the furniture is simple as each piece comes with illustrated step-by-step instructions and only simple tools are required.

Tolerances

When preparing materials and components for manufacture, each part usually has a size **tolerance**. This is the acceptable error of the dimension. For example, a piece of steel rod for a lamp needs to be 1000 mm long. If it has an acceptable tolerance of 1% then the material could be in the range of 990 mm to 1010 mm to be acceptable in a quality control check. If parts are outside their tolerance range they may not fit together. This may be very important when adding, for example, electronic components to a box. If the relevant holes are not accurate, the components will not fit or will be loose.

SKILLS ACTIVITY

Look at these prototype pen holders.

1. What tools could have been used to mark and cut out the metal parts which hold the pens?
2. Describe how you would cut the body shapes of the three pen holders.
3. Name two finishes that could be applied to the main body of the pen holder.
4. What fixing methods could be used to hold the different materials together?

KNOWLEDGE CHECK

1. What are the benefits to the user of standard components and KD fittings? (2)
2. Why are KD fittings and standard components beneficial to manufacturers? (2)
3. What is a material tolerance? (1)
4. Why is marking out so important? (2)
5. What tools and equipment would you use to mark out a sheet of aluminium? (2)

> **KEY TERMS**
>
> **datum:** the point used as a reference point for all measurements when marking out material
>
> **knock-down fittings:** fittings that can be easily put together, normally using only a screwdriver, drill, hammer or other basic tools. They are temporary joints, although many are used to permanently join together items such as cabinets and other pieces of furniture that are purchased in a flat pack.
>
> **standard components:** a pre-prepared part that is used in the production of many products
>
> **tolerance:** the acceptable margin of error in a measurement when marking out materials ready for cutting and shaping

1·4 Evaluation

1·4·1 Evaluation in design

LEARNING OBJECTIVES

By the end of this unit you should:
- ✓ be able to evaluate existing products/systems
- ✓ be able to test the performance of a product or solution against a specification
- ✓ be able to use different strategies to assess effectivenes of products
- ✓ be able to suggest possible modifications to a design.

WHAT IS EVALUATING IN DESIGN?

Evaluation takes many forms. In the study of Design & Technology at school, as well as in industry, evaluation is a continual process that takes place all the way through the design and make activity. At each stage, however, the approach and style of evaluation will differ.

When evaluating in the design process there are different styles of evaluation that can take place:

- **Formative evaluation**
  This is an ongoing and continual evaluation. The information gained can be data obtained from questionnaires and surveys or opinions. Both sets of results will help you develop your product and ensure that it meets the design specification.

- **Summative evaluation**
  This is completed at the end of the process when you have a finished product. The product is thoroughly analysed to ensure it does what was intended and, if it does not, how to improve it.

EVALUATING EXISTING PRODUCTS

Research can only start when a designer has a clear idea of what they are going to design and make, and they have a clear idea about their target market. Comparing products through the internet, magazines, and library books can be a great way of researching and evaluating products for their aesthetics. However, handling, using and physically taking apart a product allows you to gather more insightful knowledge and data. Looking at existing products is a fantastic way of evaluating how products that are already for sale are made, assembled and finished.

Disassembly

By taking apart a product you can evaluate its assembly methods, materials and aesthetics and the parts of its system. This can provide you with a great deal of knowledge that may help you when designing your own product. Taking apart a range of products provides designers with a broader knowledge of assembly methods, functionality and materials. For example, you could look at various circuits to see what are the input, process and output elements, and so on.

When looking at existing products, it is also good to look at **sustainability**. For instance, can it be **disassembled** and recycled? Are components easily replaced if broken (maintenance)?

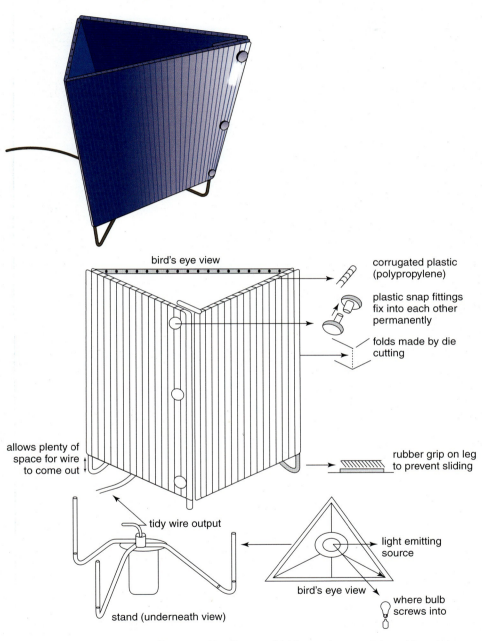

△ The lamp is very simple but effective. Taking it apart, looking at the materials and how it is assembled is a great analysis and evaluation exercise.

Similarly, if you took apart a range of breakfast cereal packages, you could investigate the materials (often carton board), types of printing process, how the net was made, the information printed on the package such as images, text, fonts, logos, nutritional data, and where the cuts, folds and glue tabs are located.

△ An example net of a cereal box

EVALUATING AGAINST THE SPECIFICATION

A design specification is a list of features a product should have. If research is thorough and the designer has a good understanding of the user and problem, the specification should be very detailed.

As a product is being developed, and once it is completed, it is very important to compare the ideas and developments to the specification to ensure it does not deviate from the original idea. Sometimes the specification points will be met and sometimes they will not. As part of the evaluation process, the designer needs to decide whether the idea needs to change or if the specification needs to be reconsidered. In this case, it is very important to work closely with the client and end user to decide if it may be possible to modify the specification and in what ways. Initially the specification needs to be followed closely, however, to ensure it meets the customer's requirements.

Evaluation

Evaluation against specification

Specification	Evaluation	Specification	Evaluation
To design and make a lamp that is sound operated.	This is covered by the circuit. The circuit has been completed and tested, and as it was in the prototype section, the lamp operated by sound.	Box is expected to appeal to teenagers, following present fashions.	The design is smart and stylish, and should appeal as a contemporary looking product. However, the colours might not appeal, but I will find this out by research.
Should be battery operated.	The circuit is operated by two batteries: one to power the circuit, and one to power the lamp.	It will be of low voltage since it is battery operated.	The voltage used is only 9V, and is powered by batteries. This should be within a safe range for home use.
Should be light enough to be portable but not too light to be unstable.	The box (including the circuit) is relatively light. It should not be any trouble for anyone to carry around. It is also not too light because it sits perfectly and stable.	Warning labels should be placed to warn of the electronic items and heat hazards and not to tamper with the circuitry/light.	I have made warning labels and they can be found on the back panel of the box.
Should be small enough to not be intrusive but big enough to be able to light a small area.	The size is not very big and I don't believe the size will disturb or disrupt the user. The small bulb should be able to light up a whole desk area.	With regular usage, the only items that possibly need replacing are the batteries and bulb.	The lamp will only be used for certain periods of time of the day. At the most should be around 9 hours of usage for a night light, so it should last without replacement for a while.
Box should be big enough to accommodate the circuit and its various components.	All the circuitry and components successfully fit onto the back of the box and is firmly held using PCB mounts. The back panels close the box.	How often the batteries will be replaced will depend on the time used.	With regular usage, the batteries should only be replaced in the time period of a few months.
Should be made to be able to be comfortably held.	It was not designed to be held, however, it was designed to be portable. I don't believe this product could cause any comfort problems.	Should be made to last as long as three years providing it is used with reasonable care.	All the materials used to make the box are all strong, such as the aluminium piece, so it should last quite a long time.
Made from safe, strong and reliable materials.	The box is made of aluminium. It is safe, strong, rigid and should hold its shape. This applies to all the other pieces included with the product.	In cases of mass production, the cost proportionally decreases.	The cost of producing 100 of this product will reasonably be lower compared to relative costs of producing one because labour and component costs would decrease in mass production.
Should have a professional looking design.	The design is smooth, simple and sophisticated. I believe the design of this product has a professional look.	Focus on using environmentally friendly materials.	I don't think I have covered this aspect of my specification very well, looking at the aspect of the biodegradability of the materials I have used.

△ The above table is an example of a student's electronic lighting system evaluation.

PRODUCT TESTING AND EVALUATING

Continual testing and evaluating should be carried out all the way through the design cycle. This means that the design is continually checked and reviewed. The same applies during your studies when you are working on projects. Some specific trials and tests are described below.

User trial and field test

A **user trial** or **field test** is when the intended final users of a product work with either a prototype or the finished product. The test is semi-controlled or scientific (looking for specific responses) and the results are collected. The design team can then evaluate the information and can help develop the product.

Questionnaires

Questionnaires can be used in market research to gather information about products or possible products. They are carried out in various ways, such as internet questionnaires, telephone conversations and face-to-face interviews. The information gathered can inform the designer about user preferences.

Commercial testing

Commercial types of testing are carried out in controlled laboratory conditions. Tests are set up to investigate how products respond to certain conditions, and the results are then carefully analysed. For example, vehicle manufacturers carry out crash tests to test how the metal will crumple around a driver. From the test, parts of a car may be strengthened. They may also carry out wind tunnel experiments to investigate aerodynamics.

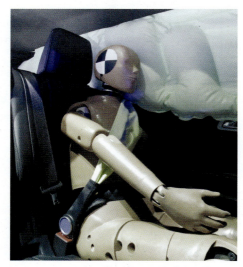

△ Commercial crash testing for vehicles uses a dummy in controlled conditions. The vehicle may be modified depending on the outcome.

Another form of commercial testing is when the product, or parts of the product, have been completed. The tests include manufacturers setting up conditions that simulate the product being used and preparing testing rigs. These are machines that will carry out a test many thousands of times to check that the product does not fail. This is sometimes called functional testing. An example may be opening and closing a lid of a scanner to test the hinges and also test the wiring and sensors. These types of tests could be carried out on any projects you work on.

Materials testing

Materials testing is also carried out in controlled laboratory conditions. The materials are subjected to specific tests such as a conductivity test to see if the material lets electricity through.

Peer review

These are tests carried out by peers, that is, classmates or colleagues. The idea is to work together to establish any faults or problems with a design in an informal way. This can be extremely useful to get honest feedback from people who have an understanding of what you are doing.

EVALUATION 2

FEEDBACK FOCUS GROUP

A FOCUS GROUP OF POTENTIAL CUSTOMER WAS ASSEMBLED TO VIEW AND ANALYSE OUR FINAL PRODUCTS OF OUR MOOD LAMPS. THEY WERE THEN REQUIRED TO CONTRIBUTE SOME FEEDBACK ABOUT WHAT THEY LIKED ABOUT THE DESIGN IDEA AND WHAT ASPECTS OF THE LAMP THEY THOUGHT NEEDED IMPROVEMENT.

1. ALL SAID THAT THEY MUCH LIKED THE IDEA OF BEING ABLE TO REARRANGE TO COLOURED PANELS TO "CREATE YOUR OWN ART!"
2. MOST OF THEM WERE BALE TO RECOGNISE THAT THE DESIGN OF THE LAMP WAS BASED ON THE ARTWORKS ON PIT MONDRIAN

IMPROVEMENTS

THE FEEDBACK FROM THE FOCUS GROUP, HOWEVER, WAS NOT CRITICAL ENOUGH ON THE FLAWS OF MY PRODUCT AND DID NOT MENTION MANY IMPROVEMENTS THAT COULD BE IMPLEMENTED TO THE LAMP'S DESIGN. AS A RESULT, I HAVE REFERRED TO THE SPECIFICATION TO OBTAIN SOME OF THE IMPROVEMENTS THAT I HAVE APPLIED TO MY PRODUCT'S NEW DESIGN [BELOW].

IN MY ORIGINAL DESIGN OF THE LAMP, THE NUMBER OF COMPONENTS BECAME A CONCERNED ASPECT AS IT RESULTED IN THE FINAL PRODUCT BEING HEAVY IN WEIGHT. TO COUNTERACT THE WEIGHT PROBLEM, BY REDESIGNING THE WAY THE COLOURED PANELS ARE CONTAINED AND ACCESSED; 2 HOLDING PANELS ARE NOT NEEDED TO CONTAIN THE COLOURED PANELS AND THE DIVIDER PANEL IS MADE OBSOLETE. NOW THE LAMP FACES ARE MADE OF SINGLE LAYER PIECES, THE LID REQUIRED OT BE THINNER AND REDUCED IN WEIGHT. I CHANGED THE AMOUNT OF MATERIAL USED FOR THE LID, TO REDUCE THE STRESS ON THE LAMP WALLS FROM THE LID'S WEIGHT. THE SAME ALTERATIONS WERE APPLIED TO THE BASE TO REDUCE THE LAMP'S OVERALL WEIGHT.

I HAVE IMPROVED THE METHOD OF REARRANGING THE COLOURED PANELS AS IT REQUIRES LESS HASSLE TO ACCESS THEM. THE LID NO LONGER NEEDS TO BE REMOVED IN ORDER TO ACCESS THE COLOURED PANEL SLOTS. THE DESIGN FEATURE IS NOW MADE ACCESSIBLE STRAIGHT FORM THE LAMP FACE; AS THE COLOURED PANELS CAN BE SLOTTED ONOT THE SINGLE LAYER OF FROSTED TRANSPARENT ACRYLIC WITH THE USE OF ROUNDED SLOT HOLES.

△ An example of a student's peer evaluation

MODIFICATIONS BASED ON EVALUATIONS

During the evaluation process it is likely that some areas of improvement will be identified, or there may be areas where the specification has not been met. These points need to be addressed, and the design team will look at various aspects of the product. A good framework and starting place for this is ACCESSFM:

- **A**esthetics: Does the product meet the aesthetic requirements of the customer and client?
- **C**ustomer: What feedback has the customer offered about the product?
- **C**ost: Has the cost of the product met the intended price?
- **E**rgonomics: Does the size of the product and its use work to expectations?
- **S**afety: Is the product safe?
- **S**ustainability and environment: Is the product sustainable and environmentally friendly?
- **F**unction: Does it work as expected?
- **M**anufacturing: Has the manufacturing process gone smoothly? How can it be improved?

Are all products perfect? Of course not: companies are constantly looking at their products to see if they can be improved for the customer. These changes are often very small, but sometimes products are redesigned due to technology and other changes that make it necessary to redevelop the product.

TOP TIP

Keep a logbook of all testing being carried out when working through the design process. It is useful as a reminder when evaluating later.

SKILLS ACTIVITY

Look carefully at a bicycle. Make a list of the tests you think the manufacturer would have carried out on the bicycle. Give reasons for your answers.

DESIGN IN ACTION

In the design and manufacturing world, the process of evaluation at every stage of the design and making process is called Continual Improvement (CI). It is an ongoing improvement process to make better products and services. The improvements come through incremental and small gains changes. Over time the products and services get better. A famous CI process is called Kaizen and was successfully implemented into Japanese manufacturing by companies like Toyota. Kaizen is credited for Japan's competitive and commercial success.

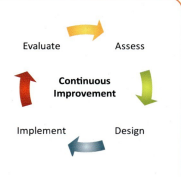

KNOWLEDGE CHECK

1. What type of technique would be best for checking the function of a prototype? (1)
2. Explain the term *peer evaluation*. (2)
3. Make a list of different methods used to test product effectiveness. (4)

REFLECTIVE LOG

Choose a product you have made recently at school. Using one or more of the evaluation techniques in this unit, work out how you would evaluate and improve the product if you were to make it again.

KEY TERMS

commercial testing: detailed and controlled testing carried out in laboratory conditions

disassembly: the taking apart of a product in order to look carefully at how it is made, the materials used and processes involved in manufacture

field test or **user trial:** a semi-controlled test by intended users of a prototype or finished product

formative evaluation: evaluation that is carried out continuously during the development of a product

summative evaluation: evaluation that is carried out on a finished product, usually resulting in a detailed report

sustainability: the ability of materials to be used widely with little or no negative impact on the environment

1·5 Health and safety

1·5·1 Safety for all

LEARNING OBJECTIVES

By the end of this unit you should:
- ✓ be able to show correct use of hand tools, machine tools and equipment
- ✓ be able to demonstrate proper regard for mandatory and other safety precautions
- ✓ understand the responsibilities of the designer to ensure that products are safe
- ✓ understand the importance of personal safety and that of others
- ✓ be able to recognise basic safety symbols used in the workshop.

WHY HEALTH AND SAFETY IS IMPORTANT

In schools, businesses, factories and places of work there is always a potential for accidents. The owners and managers of such places have a responsibility to ensure that people working in these environments are safe. In most countries there are important health and safety laws, legislation and guidelines to protect employers and employees (workers) in the workplace. Health and safety organisations are responsible for the encouragement, regulation and enforcement of workplace health, safety and welfare. They impose criminal liability on owners of businesses that do not meet regulations. However, the levels of safe working practice across the world vary considerably. This may be due to limited laws or legislation, or lack of enforcement by relevant authorities.

Most schools will have a set of health and safety workshop rules and regulations. These are put in place to make the school workshop a safe place to work, be creative and make products. They will differ from school to school but essentially assess the risks for each environment and control the risks with a set of rules.

ACCIDENTS AT WORK

Most accidents in the work environment are caused by carelessness and simple human error, but having an accident or injury in a work environment can have various consequences:

Employee:
- injury that could be temporary or permanent
- time off work, possibly unpaid, which would have further implications for the family of the victim.

Employer:
- the need to find replacement staff
- having to replace damaged equipment
- legal costs and compensation to the injured.

It is important to take all necessary precautions in order to reduce risk of accident and injury at all times.

SKILLS ACTIVITY

Health and safety is a serious issue for business and industry. Consider what other effects injuries have on employers and employees.

RISK ASSESSMENT

Carrying out a **risk assessment** for activities is very important in reducing and managing accidents. Every school has a member of staff responsible for this. In your Design & Technology class, your teachers will ensure you work in a safe environment. Hazardous activities will have procedures and/or equipment in place to reduce the risk of injury.

Managing your working environment is very important in reducing the risk of an accident. Below are some points that should be considered in Design & Technology classrooms, which also apply to industrial environments.

The list below sets out some of the general risk assessments carried out in a Design & Technology classroom and in industry:

- size of the room and the number of students
- storage areas for chemicals
- storage areas for tools, equipment and materials
- space for students, teachers and technicians to walk around
- emergency stop buttons to cut the power to machines
- fire extinguishers
- emergency exits
- barriers around machines
- lines painted on the floor to indicate areas students can and cannot go
- adequate lighting, ventilation and working temperature
- relevant signage to inform of potential hazards and dangers
- **Personal Protective Equipment (PPE)**
- first aid equipment
- waste disposal areas.

More specific risk assessments will apply to the many different activities that take place.

SKILLS ACTIVITY

Think of a range of specific tasks you may undertake in your Design & Technology classroom, and list the appropriate risk assessments and actions that you may consider.

In addition, it is also important to keep all working areas clean and tidy, put away tools and equipment after use, keep worktops, floors and stairs clear by vacuuming and sweeping to reduce dust, and to avoid surfaces becoming slippery.

Schools usually have rules for the workshop or practical making environment. Specific rules for your school should be shared with you at the beginning of your course. An example of some general safety rules are:

- do not run in the workshop – you may trip and fall on or into machinery
- wear appropriate PPE to protect yourself from machinery and hazardous substances
- tidy away and clean up all resources after use, so that the workspace is clean and free from any potential hazards
- place bags and equipment not required in the lesson in a designated storage area, to keep walkways and public spaces clear so you do not trip over anything
- know the location of emergency exits, fire extinguishers, the first aid kit and emergency stops, so you can act quickly and appropriately in case of an emergency
- ask a teacher if you are unsure of what to do.

When working in a practical environment you will need to demonstrate that you can be trusted to work safely. To do this, you should be aware of potential dangers, assess risk, and seek guidance where required. Before using equipment and machines, ensure you understand how it should be used safely. You may require a demonstration of how it should be used, including any safety features. If you are unsure, always ask!

PERSONAL PROTECTIVE EQUIPMENT (PPE)

Using Personal Protective Equipment is a simple and effective way of protecting yourself in a practical environment. Goggles, aprons/overalls, dust masks, visors, sensible footwear and ear defenders are all useful pieces of equipment depending on the activity taking place. Please see the table of symbols on the next page.

Some tasks that are undertaken may require extra precautions, PPE and safety considerations:

- when soldering you may require extra ventilation and extraction to remove smoke and fumes that is produced directly where you are working
- when finishing materials you may need ventilation and extraction to expel fumes
- when using particular solvents and adhesives you may require gloves and additional ventilation and extraction to protect your skin, and to prevent inhalation of fumes.

Workshop operation	Hazard/Risk	Suitable PPE	Sign/Symbol
General practical activity in the Design and Technology environment	Protection of clothing from dust, paint, chemicals	Apron or overall	
Drilling, wasting, sanding, polishing	Dust, sparks, particles in your eyes	Goggles/visor	
Handling hot or sharp materials Using adhesives	Potential to burn fingers Cutting hands when handling materials Solvents on skin	Leather gloves / disposable gloves	
Drilling, wasting, sanding, polishing	Breathing in dust, sparks, particles	Dust mask	
General workshop noise, specific noisy task (grinding)	Ear damage from prolonged exposure to noise	Ear defenders	

HEALTH AND SAFETY REGULATION AND GUIDANCE

There are a number of different organisations across the world that provide health and safety guidance. They all will have similar objectives, but the specific regulations and legislation will vary from country to country. Their main objectives are as follows:

- To secure the health, welfare and safety of people in any place of work
- To protect people against health and safety risks arising from work activities
- To control the use of chemicals, corrosives, flammable and dangerous substances
- To control emissions released into the atmosphere through air and water.

Here is a list of some of the organisations around the world that offer guidance on health and safety, and that aim to protect people and the environment:

- British Standards Institute (BSI)
- Design and Technology Association (DATA)
- European Agency for Safety and Health at Work (EU-OSHA)
- International Labour Organization (ILO)
- International Atomic Energy Agency (IAEA)
- Occupational Safety and Health Administration (OSHA) in America.

WORKSHOP SAFETY SYMBOLS AND SIGNAGE

As well as the symbols for PPE, there are many other safety symbols that you may see in a workshop environment.

Some are very general signs but many relate to a set of standards called **COSHH: Control Of Substances Hazardous to Health**. COSHH is the regulation that requires employers to control substances that are hazardous to health. This is a UK-based set of regulations, but many other countries use COSHH or have similar standards. The COSHH regulations give advice on how to handle and safely dispose of chemicals, solvents and other potentially hazardous materials.

△ An emergency stop button   △ A sign for a fire extinguisher   △ A First Aid symbol

COSHH symbols and signage

Products with labels like the ones below must be stored in a lockable area in a metal cabinet, with clear labelling that the contents are potentially hazardous.

△ corrosive   △ flammable   △ toxic   △ harmful

SKILLS ACTIVITY

Health and safety signs are usually placed in workshops. Can you find any in your workspace? Identify the safety signs and explain what they mean.

HEALTH AND SAFETY IN DESIGN & TECHNOLOGY

The Design & Technology areas that are available in school vary considerably. However, all workshop spaces have the potential for risky activity. Hand tools and small machine tools can be hazardous if not used and stored properly. Remember to follow the correct working procedures and ask for a demonstration if you are unsure.

HEALTH AND SAFETY AND THE DESIGNER

We live in a globalised world where, for example, products may be designed in Europe but made in South America by a different company. This is called **out-sourcing**. Another practice is to move a company's manufacturing facilities overseas or to another manufacturing company. This is called **off-shoring**. Companies do this to take advantage of lower labour costs, a larger supply of workers and increased profits. When products are outsourced or sent off-shore it is more difficult to control the other companies' workers and practices in terms of both health and safety and environment. Designers can help by ensuring the design and manufacture of products will be sustainable and help ensure they are safe to produce.

Organisations such as the British Standards Institute produce guidance and tests for products to ensure they are safe. For example, the regulation (or 'standard') BS EN 1022 relates to the stability of chairs. BS EN 71 looks at the safety of toys.

To ensure products are safe for their market, the standards above (and many others) are used by designers and their companies to produce quality products. The BSI website has an education section that is extremely useful.

Health and safety is extremely important at school, in a design studio and in industry, during both large- and small-scale manufacture. When designers are creating ideas and solving problems it is important that they consider health and safety and the environment at all stages of a product's life, including its manufacture, as well as its use and eventual disposal.

Designers have the responsibility of ensuring:

- products are safe to use when purchased, that they will not fail or cause injury, by putting them through suitable safety tests to meet safety standards such as those set out by BSI and CE, a marking which ensures the product complies with EU safety and legislation and can be sold throughout the EEA

- products are correctly labelled and that any safety warnings are present on products and packaging
- thorough Quality Assurance and Quality Control checks have been implemented during the manufacture of the product.

Some important health and safety questions you need to consider when designing include:

- are the materials used toxic for the people manufacturing the product, the people using the product, and when it may be thrown away into landfill?
- is the product safe for customers to use? For example, is there a risk of potential electrical shocks, or small parts that could be swallowed by a child?
- will the product biodegrade or will it corrode and contaminate the environment when disposed of, for example, batteries (included in many products)?

Designers are in a position to choose materials and processes that can be very safe for manufacture and long-term use, and which also help the environment. For example, designers in Asia may use bamboo instead of hardwood for furniture production, as this is more sustainable.

TOP TIP

When you are involved in the designing and making of products, your teachers will monitor your ability to work correctly and safely with hand tools, power tools and machines. Safe working practice and being responsible in a potentially dangerous environment is very important, so always keep this in mind.

SKILLS ACTIVITY

Look around your Design & Technology classroom at school. Make a list of at least 10 potential health and safety hazards, and the risk assessments you would put in place to avoid any accidents.

Volvo, the car manufacturer, has always been very innovative when it comes to safety. They introduced the seatbelt in 1959.

Research three other health and safety innovations that they are credited with.

KNOWLEDGE CHECK

1. Describe two safety factors that need to be considered when using adhesives to join two pieces of acrylic together. (2)
2. List five items of PPE. (5)
3. Explain what PPE and safety precautions you should consider when soldering an electronic circuit. (2)
4. How can a designer influence the health and safety of workers in a factory? (4)
5. In school, why are health and safety rules put in place? (2)

KEY TERMS

Control of Substances Hazardous to Health (COSHH): a set of regulations designed to ensure the safe control of hazardous substances

off-shoring: the movement of a company's operations, such as manufacturing or commerce, to another country, often to save costs

out-sourcing: the use of a third party contractor to carry out work and supply services

Personal Protective Equipment (PPE): clothing and equipment, such as goggles or helmets, designed to protect a person working in a making environment

risk assessment: an examination of the possible risks in an area and a statement of how they should be managed

1·6 Use of technology

1·6·1 Use of technology in designing and making

LEARNING OBJECTIVES

By the end of this unit you should:
- ✓ understand the need to research existing products (internet)
- ✓ understand the benefits of CAD/CAM in different production systems
- ✓ understand how CAD can be used to generate 2D and 3D images
- ✓ understand how CAD/CAM is used in industry
- ✓ be aware of the variety of machines that are computer controlled
- ✓ understand how computers can enhance stock and quality control.

TECHNOLOGY IN DESIGNING AND MAKING

Technology is central to the design and making process. From defining a problem through to evaluating a solution, technology and computers are involved. Here are some examples of how technology can be used in the early stages of designing:

- interviews can be recorded using microphones
- results can be analysed and presented in graphs, tables and presentations
- drawings can be created on tablet PCs
- research can be carried out via online forums
- questionnaires can be carried out through social media
- existing products can be evaluated over the internet.

CAD AND CAM IN MAKING AND MANUFACTURING

Computer-aided design (CAD) is an integral part of the design process in most manufacturing situations today. CAD is usually central to design development and manufacturing and offers many advantages, such as:

- the ability to draw in 2D and 3D and then change a design quickly without redrawing from the beginning
- allowing the designer to see a drawing from different angles by rotating and zooming
- being able to render a drawing to make it look photorealistic and in 3D
- the ability to send a design to a machine to make individual parts using **computer-aided manufacture (CAM)**
- the ability to test a design, such as stress and strain, through testing software before manufacture, saving money on expensive models
- managing production schedules through CAD to control work flow through a factory
- carrying out Quality Control, so that goods are produced to the right standards.

In terms of product development, CAD can be two dimensional (2D) or three-dimensional (3D). Two dimensional CAD packages allow the user to produce accurately scaled, dimensioned technical drawings that can, if required, include text, photographs, bitmap images and vector graphics. This is very useful in school for producing designs such as logos, **point of sale displays**, **packaging nets**, menus **and PCB circuit layouts**. Two dimensional CAD drawings can also be output to a variety of CAM devices.

CAD software includes programs such as TechSoft 2D design, CorelDRAW and Sketch Up, often used in schools, but other more industrial software may be available such as the software from Autodesk, SolidWorks and PTC Creo.

△ A professional CAD rendering

PRODUCTION SYSTEMS

There are a number of different production systems used by artisans, makers, small businesses and the design industry. The type of system will depend on the product and the business.

One-off production, job production, bespoke production

This type of production is where only one product is made at a time. It is highly skilled and often labour intensive, and the product is made by very well-trained people. One-off jobs are usually expensive. The clientele often requires a unique piece of individual high quality, such as handmade furniture, jewellery or individually engineered items.

Batch production

This is where small quantities of the same product are made. It can be labour intensive but the use of machines, some automation and the use of templates, **jigs**, moulds, patterns and formers, aid and simplify repetitive tasks and speed up production. Batches can change easily so the products can be modified. While sometimes expensive, depending on the product, the cost per individual item is reduced because of the larger quantity made and the speeding up of the process.

Mass production

This is when hundreds and often thousands of the same product are manufactured. The process is often automated. Individual parts can be assembled into products. There are often fewer workers in this process as many machines will undertake many of the operations. This type of production is very expensive to set up because of the technology and machinery costs. Lots of **quality control (QC)** checks need to take place to ensure that all the products meet standards; for example, checking to see if the print colours in a package design are correct and not faded, or checking the dimensions of a component so it is guaranteed to fit.

Continuous production

Continuous production is similar to mass production but production never stops. Making is often 24 hours a day with people working in shifts to support the production.

An example of continuous manufacturing is the production of fuels, oils, plastics and lubricants from crude oil. The petrochemical industries will continually refine the crude oil to produce the products for sale.

CAD in industry

Computer-aided design is used in all production systems. Free software is often available to individuals and small companies. More sophisticated, higher quality software is available to purchase. CAD is often used in the initial design stages to see a 3D image of a product and view photorealistic images of designs. Designs can then be sent to 3D printers, which can make prototypes. A 3D drawing file is usually converted to a stereo lithography file (.stl) for this purpose. Two-dimensional CAD drawings or sub-assembly drawings can be sent to Computer Numerical Control (CNC) routers, laser cutters, etc., to manufacture parts.

Remote manufacturing

In the global society we live in today, designing may take place in one country and manufacturing may take place in another. People may never meet and all correspondence may be done electronically.

Computer-aided manufacture

CAM is where a 2D or 3D CAD drawing is converted into co-ordinates. The machine head or cutting element connected to the CAM machine then goes to the co-ordinates to complete its operation, which may include cutting with a tool, laser or knife. For 2D the machine will follow X and Y and for 3D, X, Y and Z paths. CNC often describes this type of machine.

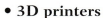

△ XYZ co-ordinates used in CAD and CAM

Types of CAM

There are many types of CAM machine available, from entry level educational machines to multimillion dollar industrial machines. Some of the more common CAM machines available in schools are:

- **Vinyl cutter**
  Vinyl (plastic sheet with adhesive backing, like stickers) sheets are fed in and out of the machine with rollers and the tool head moves left and right. The combination of this movement and a blade to cut the vinyl allows 2D shapes to be cut.

- **Laser cutters**
  Laser cutters use a laser beam to cut 2D shapes in a wide range of materials, including leather, laser plywood, paper, card, acrylic, fabric and many more. The laser cutter can also etch into materials at various depths to give a 3D look.

- **3D printers**
  These printers have been around for many years but recently have become available as desktop machines. They are used to create **rapid protoypes** of products and can also be used to create working components. There are a number of systems available which work in slightly different ways. The most common in schools is additive manufacturing where a material is extruded through a nozzle and the component is built up slowly in layers.

- **Computer Numerical Control (CNC) routers and millers**
  CNC routers and millers are more industrial and a tool head is used to cut out, cut slots and many other operations. They work on materials such as aluminium, steel hardwoods and manmade boards.

△ A vinyl cutter

△ A laser cutter

Other more industrial machines may include waterjet cutters and plasma cutters.

Many modern manufacturing companies will use a range of CNC machines to make the components of their products. After components are made, they are then assembled on a production line. **Computer-integrated Manufacture (CIM)** is where computers are central to the whole process and monitor manufacturing, stock control, quality and output.

> **TOP TIP**
>
> The use of CAD/CAM is encouraged in this course. If facilities are available, make yourself familiar with both the software and hardware and utilise these skills in your project.

COMPUTERS IN QUALITY AND STOCK CONTROL

Computers are used heavily in commercial activities to control stock of the products being sold and also to monitor the quality of the products. In manufacturing and distribution, radio-frequency identification (RFID) devices are now commonly used. RFID are wireless non-contact devices that use electromagnetic fields to transfer data, for the purposes of automatically identifying and tracking tags attached to objects.

△ A barcode reader

Barcodes, the stripy lines that can be found on the packaging of most products are used extensively from manufacture to distribution and supply to the customer. They are usually seen in a supermarket to monitor the quantity of products being sold, and to enable the cashier to scan each item to register the price. However, the life of a barcode begins much earlier than when it was first stamped onto a product.

Each barcode has a lot of detail built into its individual number, such as the country where the product was made, the type of product (classification), and date of manufacture. Some barcodes are more pictorial to make them more interesting. There are many types of barcodes used in different industries.

Barcodes in action

When a product is being manufactured, many of the small components and sub assemblies (smaller groups of components that make up a larger product) will have a barcode. When each piece is used, the barcode is scanned and this tells the warehouse the quantity of parts on the assembly line. This will inform the stock controller when to supply more parts so they do not run out. When a product is finally assembled, the whole product will be given another barcode. The product will be sent for distrubution to the customer at the retail outlets. The retailer scans the code at each sale, again to inform their own stock control.

△ A barcode developed specifically for a product

The barcode can also be used later if there is a problem with a product. If a product fails within its warranty, the barcode could be used to help the manufacturer track and investigate the root of the problem at the manufacturing stage and improve their quality control.

SKILLS ACTIVITY

Pick an object in the room you are working in. What type of production process would have been used to make it: mass, batch or one-off production? Explain why.

Where would CAD have been used in the process?

> **DESIGN IN ACTION**
>
> This book you are reading is an example of remote design and manufacturing. The book was written in the UK, Brunei and Hong Kong. The editing was completed in the UK, typesetting in India and the manufacturing in China. All correspondence was completed by email, cloud computing, online conference calls and telephone. Distribution is worldwide.

KNOWLEDGE CHECK

The use of CNC machines in industry is very common.
1. What are four advantages and four disadvantages of CNC machining? (8)
2. How does rapid prototyping help a designer and design team? (2)
3. What are the advantages of CAD over traditional drawing methods? (4)
4. How can barcodes help with quality control of products? (4)

KEY TERMS

Computer-aided design (CAD): the use of computers to design products on a screen. Also used in production management.

Computer-aided manufacture (CAM): the use of computers to automate the manufacture of products and to control production processes

Computer-integrated Manufacturing (CIM): a method of production in which design, manufacture and management are linked electronically and controlled by computers

Flexible Manufacturing Systems (FMS): manufacturing systems that produce products and goods using robots and computer-controlled tools and machines, without human intervention. The products can be changed easily by using a different program.

jigs (also templates, patterns and moulds): simple devices that speed up marking out and repetitive tasks

packaging net: a two-dimensional piece of material such as cardboard, designed to be folded and fastened into a three-dimensional piece of packaging

PCB circuit layout: a line drawing that represents the conductive tracks that join components in an electronic circuit

point of sale display: a board or other form of visual advertising that is displayed at the place where a product is purchased

quality control: the process of maintaining standards in manufactured products

rapid prototyping: use of CAM to quickly produce models or parts for testing and modelling

1·6·2 Systems

LEARNING OBJECTIVES
By the end of this unit you should:
✓ be able to identify the features of a control system
✓ understand the features of a control system and applications.

WHAT IS A SYSTEM?

A system is a combined group of individual components that work together to perform a task or activity. Systems can be in many forms and include:

- electronic systems
- pneumatic systems (compressed air)
- mechanical systems
- electrical systems
- hydraulic systems.

A system could also be a combination of two or more of the above depending on the task to take place. All of the separate systems are called sub-systems. For example, on a building site there may be an excavator. This machine will use mechanical, hydraulic and electrical systems to perform its primary function of digging and earth-moving. However, it will also have secondary sub-systems such as lighting, braking, sound and heating/cooling. All of these systems work together seamlessly.

CONTROL SYSTEMS DIAGRAM

Systems are often broken down and analysed in a diagrammatic form to make them easier to understand. This is often called a systems diagram. The most basic systems diagram is an open-loop system, which comprises an **INPUT**, an **OPERATION** (sometimes called CONTROL) and an **OUTPUT**.

△ Open-loop system diagram

Each part of the system has a function and this may vary between the types of system. For example:

- The input could be mechanical movement, electrical switches or movement sensors. When the Input changes it will pass on this information to the Operation stage of the diagram.
- The Operation stage takes the signal or movement from the Input stage, processes it in some form, and then passes the new information or signal to the Output stage.
- The Output stage reacts to the Input and Operation and responds accordingly. This could be movement or a light coming on and off, for instance.

For example, the brakes of a car:
- The driver pushes down on the brake pedal. There is pressure on a mechanical lever (INPUT).
- The lever activates hydraulic fluid in the pipes of the braking system, which pushes a piston (OPERATION).
- The piston pushes on the brake, which squeezes to slow the wheels (OUTPUT).

This type of system has no control and the output only reacts to an input. The **feedback** is human. Many systems control themselves and incorporate FEEDBACK to pass information from the output to the input. This is called a closed-loop system. The anti-lock breaking system (ABS) in a car is a good example of this.

△ Closed-loop system diagram

The FEEDBACK link closes the open loop and sends signals from the OUTPUT to the INPUT.

For example, an air conditioning system:
- When an air conditioner is turned on (INPUT) it is usually set to a certain temperature, for example 21° Celsius (70° Farenheit).
- The cooling system will cool a room (OUTPUT) and a thermostat will monitor the temperature (FEEDBACK). When the room temperature cools enough and reaches its set temperature it will switch off until it gets warm again, at which point the thermostat will FEEDBACK to the INPUT to switch on again. This process is continual and keeps the room at a constant temperature.

 REFLECTIVE LOG Consider a modern mobile device or tablet. How different is the system to conventional electronic systems?

TOP TIP

When designing a project that is made up of various different systems, break it down into blocks in a systems diagram to make it easier to understand.

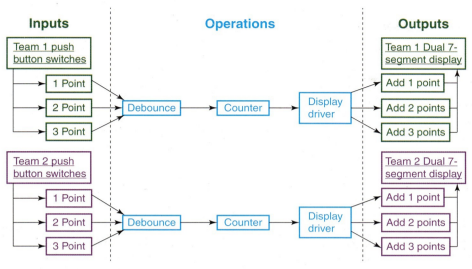

△ A sports scoring system diagram

SKILLS ACTIVITY

Draw a systems diagram for a product you are familiar with, such as a bicycle. What are the inputs, controls and outputs of this product?

KNOWLEDGE CHECK

1. Match the following systems:
   - a steam train
   - a telephone
   - earth-moving equipment
   - changing a Formula 1 racing car's wheels
   - hydraulic
   - pneumatic
   - mechanical
   - electronic   (4)

2. Give one more example of each of the following types of systems:
   - mechanical system
   - pneumatic system
   - electronic system
   - hydraulic system   (8)

> **KEY TERMS**
>
> **feedback:** the use of sensors to read the output and tell the input what to do
>
> **input:** the signal that starts a system
>
> **operation (*or* control):** the 'brain' of a system. It takes an input, decides what to do and sends an appropriate message to the output.
>
> **output:** a response to the information from the process element of the system

1·7 Design & Technology in society

1·7·1 Design in society

> **LEARNING OBJECTIVES**
>
> By the end of this unit you should:
> - ✓ understand the role of designers, artisans and technologists in industry and society
> - ✓ understand the effects of Design & Technology activity on social, moral and economic issues
> - ✓ understand a range of human factors that need to be considered when designing.

THE ROLE OF THE DESIGNER

We are surrounded by design. You may be sitting in a classroom, a library, or at home studying. You may be sitting on a chair, reading at a desk, listening to a mobile device with music playing, wearing a range of garments and jewellery. All around you there will be objects that have been designed and made. There are so many that you probably do not really think about it! Even this book has been designed and manufactured.

Designers, **artisans** and technologists have similar roles in that they play an important part in creating products, solving problems and coming up with solutions. An artisan may be a very skilled craftsperson such as a furniture maker or jewellery designer. A technologist may support engineers and designers with creative solutions to technical aspects of a design, such as electronic circuitry or mechanical systems.

Designers, artisans and technologists are people that may work individually but, especially in larger companies, they often work together in teams to make products, environments and systems, such as web design, architecture, industrial design, fashion design and engineering. They will generally follow a design process similar to that discussed in units 1·1 to 1·4. Their role is to create products that we can enjoy using and that will, hopefully, make our lives easier.

A designer will initially take the design brief from the client, then immerse themselves into the design process by carrying out a range of research activities and asking questions, such as:

- Who is the user of the product?
- What are their general interests and habits?
- Are there size constraints that need to be thought about?
- What environmental opportunities are there?
- How can the design help to improve sustainability?
- Where will the design be used?

- How will the user use the product?
- Are there existing products available? How can they be improved?

Once all this information has been gathered and analysed the designer or design team should have a very good idea of what is required. They can then write a design specification (see unit 1·1·2) and start the fun and creative process of coming up with new innovative ideas to solve the problem.

SKILLS ACTIVITY

Can you make a list of very highly skilled artisans? For example, a wood carver.

EFFECTS OF DESIGN & TECHNOLOGY ACTIVITY AND WHY A DESIGNER MUST THINK SUSTAINABLY

The role of the designer is a very important one. Designers have social, environmental and economic responsibilities. At each stage of the process, the designer and/or the design team must consider the impact of their design decisions on other people, the environment we live in and the economy. Sustainability should be at the centre of the designer's thinking. As well as carefully considering the design and manufacture of products, designers should also think about what happens to a product when it is broken and not needed anymore. **Life cycle analysis** and philosophies such as '**cradle to cradle**' thinking need to be taken into account, so that the whole process of designing, making and what happens to the product when it is finished with are considered carefully.

The Venn diagram below shows economic, social and environmental issues as being closely interlinked. A truly sustainable product will consider all of these elements and be at the core of the diagram.

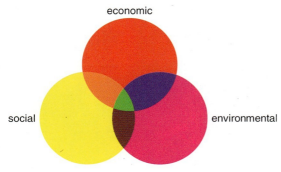

△ A Venn diagram demonstrating a truly sustainable product converging in the middle (the green area).

Here are some points designers should consider when designing sustainably.

Social

Ensure the quality of other people's lives and our own are not reduced by:
- avoiding exploitation of workers and traders
- sourcing materials local to the manufacturing area
- making sure the working conditions of workers are safe and of a reasonable standard
- using Fairtrade principles.

Environment

Ensure products do not exploit the Earth's resources by:
- selecting materials from sustainable sources
- making sure processes for manufacture do not cause pollution and toxicity
- making efficient use of energy during manufacturing, distribution and use
- making sure products can be recycled and/or reused.

Economic

Ensure there is a local and regional benefit to designing a product by:
- creating jobs
- making sure profits are shared fairly (is anyone being exploited?)
- making sure production creates and maintains skills.

These considerations are often very hard for a designer to take into account, especially if they are working for a large multinational company. However, if a company is actively engaged in sustainability it helps considerably. Many companies try very hard to balance economic, social and environmental issues to become more sustainable. However, the pressure and reality of business, making profits and pleasing shareholders sometimes takes over. While there have been great gains in recent years with companies being more aware of their role in sustainability, there is still a great deal of work to do.

> **DESIGN IN ACTION**
> Look up the sports shoe company Puma's 'Clever Little Bag'. This is an excellent example of a multinational company embracing sustainability. The design of the footwear packaging made production and transport easier and allowed the packaging materials to be reused.

KNOWLEDGE CHECK

1. Name a company or organisation that you feel has a strong sustainability policy and explain why. (6)
2. There are three R's we often use when thinking of the environment. What are they? (3)

SKILLS ACTIVITY

Pick a product that may be near you. Discuss with a partner the social, moral and economic issues that may be involved in producing the product.

CONSIDERING HUMAN FACTORS IN DESIGN

When creating new products designers have to take into account factors that are related to the end user. There are a number of areas to consider:

Ergonomics

Ergonomics is the study of the way end users interact with products and systems, such as how they hold them, wear them, or use any specific features.

Anthropometrics

This is the study of human sizes and measurements and how they relate to products. If you sat on a chair in a classroom that was too small for the desk, you would soon become uncomfortable and would not be able to work efficiently. Scientists have measured the dimensions of thousands of people across the world from different races and cultures. They have produced extensive mathematical data sets that can be used by designers and engineers to help produce products that fit the users.

△ In this diagram, anthropometrics data will be available for the dimensions A–E.

This graph demonstrates the heights of people. The 50th percentile is the average. 0–5 percent are very small and 95–100 percent are very tall. Designers design between the 5th and 95th percentile to meet most people's needs.

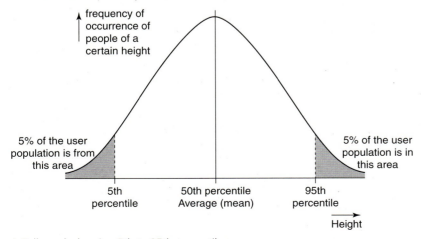

△ Bell graph showing 5th to 95th percentile

> **TOP TIP**
>
> Undertaking your own anthropometric research can be very useful but time consuming. There is a great deal of existing research and data that you can use from books and the internet, which you can analyse and evaluate for your needs. Remember to reference where you have taken the information from.

Sociological study

This is the study of how end users relate to their environment and with each other. The way we interact in and around products, devices and services is important for our personal well-being and comfort. For example, the spacing of seating in public places in restaurants, and the distance between the aisles in a supermarket, are things designers need to consider to make the experience more user-friendly. Designers also need to consider wide ranges of human needs to accommodate a broad range of people, for example wheelchair access and braille.

Physiological study

This is the study of how products affect physical comfort and the movement of users. The way people move and co-ordinate their bodies when carrying out tasks can sometimes cause discomfort and create safety issues. For example, the driver of a bus will sit in the driving seat for many hours and perform many tasks during their job. The location of pedals, switches, levers, mirrors, etc., are very important for the comfort and safety of drivers and their passengers. Many end users may have genuine physical problems that do not allow them to use a product properly or efficiently. For example, turning on a tap would be difficult for people with limited movement in their hands, so designers help to produce devices to support people with such problems.

△ Common products designed to fit the hand

Psychological study

This is the study of the way our minds interact and relate to devices and products. Our senses are very important. Touch, taste, hearing, smell and sight all form part of the way we interact with products. The way people interact with machines, products and environments can create a good or a bad experience. For example, the use of colour in traffic lights indicates a psychological instruction to stop and go, and advanced systems such as mobile devices with interactive touch screens are very common and make interaction with smartphones and hand-held electronic devices much easier, more efficient and intuitive. Interaction with products through voice recognition and movement is developing fast and will become more mainstream in the future. Google Glasses, which recognise retina movements in your eyes, is a good example of this. The relationship between end users and devices will become more user-friendly as technology develops and the status of good design, and demand for it, increases.

Awareness of human needs is very important for designers and design students. In your own designs, consider everyone who may possibly use the product and their needs.

Interaction design

Interaction design is where designers incorporate both software and hardware design into products to satisfy the needs of the majority of users. Ergonomics, including anthropometrics, aesthetics, physiology and psychology are all part of it. Look up the Bradley Timepiece – this is a watch that uses touch to help users tell the time.

SKILLS ACTIVITY

Put yourself in the position of a person with some personal limitations, such as difficulty opening a glass jar or using a tin opener when you have a plaster cast on your arm after a sport accident. Think about these questions:

- Would this simple task be difficult?
- What particular movements would you find difficult?

When designing for others, put yourself in their position to get a better understanding of their needs.

KNOWLEDGE CHECK

1. Students study in many different ways and some have nowhere to work comfortably. If you were to design a study unit, list four points you would consider to be important in terms of ergonomics. (4)
2. Explain the 5th to 95th percentile. (4)
3. Give two examples of considerations you would make when taking into account physiological and psychological design. (4)

KEY TERMS

artisan: a highly skilled person who generally produces bespoke or small numbers of high quality products or items

cradle to cradle: describes the whole life of a product from raw materials to end of life and recycling

ergonomics: the study of the way products can be designed so that they produce the best experience for users

life cycle analysis: an analysis of each part of the product's life and journey, from creation to the end of the product's life

planned obsolescence: the deliberate creation of a product with a limited life

1·8 Product design application

1·8·1 Meeting the needs of users

LEARNING OBJECTIVES

By the end of this unit you should:
✓ understand how existing products meet the needs of the users.

THE IMPORTANCE OF CONSIDERING HUMAN FACTORS IN DESIGN

Human factors is the discipline of optimising human performance to make better products and services, also known as ergonomics. It is a designer's job to ensure that the everyday tasks we undertake with the products in our environment are as easy and comfortable as possible. If a product is uncomfortable to use it will also be less efficient to use.

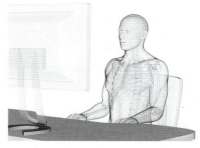

△ An ergonome at a desk

MEETING THE NEEDS OF USERS

In order to make products better, designers need to consider how they are used and how people will interact with them. Take mobile phones, for example. How do you use your mobile phone and how do you interact with it? Consider how this has changed in the last five years.

Interaction design is a relatively new field of design that looks specifically at the links between software and hardware in products. Most mobile phone devices now have large interactive screens, which use hardware and software to communicate with users. Interactive design looks at how this relationship can develop and improve.

This diagram shows some of the factors designers need to think about when considering human factors. Can you add any others?

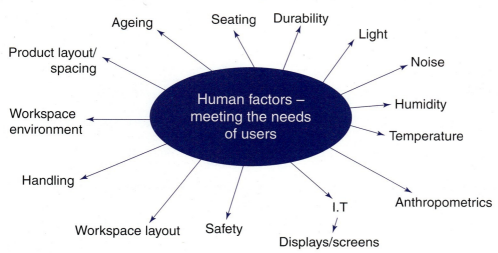

TOP TIP

It is essential that designers work closely with the users of a product to really empathise with their situation, in order to create products that really work. Always think carefully about who will be using your designs and how you can make their experience as good as possible.

SKILLS ACTIVITY

Look at the images below. Who are the users of these products and what are their needs? What do designers need to consider when developing products for these users?

△ A wheelchair

△ An aeroplane cockpit

△ Toothbrushes

△ Glasses

△ An office work station

△ A hearing aid

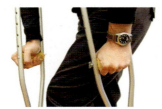

△ Crutches

KNOWLEDGE CHECK

1. Describe briefly four activities that you are familiar with that could be improved by applying good interaction design. (8)

> **KEY TERMS**
>
> **human factors:** the ways in which human users interact with products or systems
>
> **interaction design:** the aspect of a design that considers the relationship between hardware, software and the needs of the product's user

1·8·2 Considering production manufacturing

> **LEARNING OBJECTIVES**
> By the end of this unit you should:
> ✓ understand one-off, batch and mass production techniques.

PRODUCTION METHOD

As part of the design process, the designer needs to think about how many items are going to be manufactured (scales of production). This will depend on the demand. The quantity required will determine the materials and processes needed to make the product.

One-off production

One-offs are products that are designed and made for a specific use or situation that will only happen once. Most one-offs are generally handmade using a wide range of techniques and equipment.

For example, the Bloodhound SSC car, designed and built to break the world land speed record, is a one-off piece of engineering, so the design team and manufacturers only need to make a small quantity of each of the **bespoke** specialised parts. The engineers and designers are highly skilled and specialists in their roles.

The product you will be designing and making for your project will be considered a one-off. Other examples of one-offs include a custom-made bicycle for a specific competition, a custom-made wedding outfit or an exclusive piece of jewellery.

Batch production

Batch production is when a product is manufactured in a batch of a specific quantity. A batch can range from a few items to many thousands. Batch production usually takes place over a series of workstations, and what happens at one workstation can be changed to create a slightly different batch. For example, a baker making 100 loaves of bread might make a batch of 50 wholemeal loaves and 50 ciabatta for sale in a shop.

The use of jigs, moulds and templates speed up the process of making batches. For example, a jig may be used to bend a material around to keep a consistent shape when making a metal chair leg. The baker above would use a mould to put the dough into so that the loaves are the same size and shape each time they are produced.

Another example of batch production would be a monthly fashion magazine: the designers would amend an existing design template to include the new text each month and the printer would manufacture a specific number of copies for each issue depending on the size of the readership.

Mass production

Mass production is when identical products are made in huge volumes. This could range from a thousand up to millions, depending on the product. For example, standard metric nuts and bolts would be made in high volumes for sale to a huge range of businesses. Specialist machines, tailored to the specific product, are used for mass production. These machines are usually expensive but the cost of creating these machines is offset by the large volume of products that will be made and sold.

The designers of a ball point pen will be planning to make millions of each pen, and they will be mass produced using polymers that are injection moulded. Pens may be manufactured in colour batches of many hundreds of thousands.

Continuous production

Continuous production is when products are made for weeks and even months on end, 24 hours a day, seven days a week. This type of production is heavily automated. The product range is typically small and specialised.

An example of continuous manufacturing is the production of fuels, oils, plastics and lubricants from crude oil. The petrochemical industries will continually refine the crude oil to produce the products for sale.

When designing, it is important to consider the commercial processing that will take place, as this will influence the materials, form and overall aesthetic of a product. Failure to do this will sometimes mean that something is designed on paper to be made out of specific materials, but in practice would not work when manufactured commercially.

The table below illustrates the increasing or decreasing effect that various production methods have on certain aspects of manufacturing products, such as cost and efficiency.

Level of production	Efficiency of process	Labour costs	Workers' skills	Equipment, machines and tooling costs	Product costs
One off/bespoke	↓	↑	↑	↓	↑
Batch					
Mass					
Continual					

SKILLS ACTIVITY

Look at these products. Discuss with a partner how you think each is manufactured.

△ A set of disposable biro pens

△ A stack of magazines

KNOWLEDGE CHECK

1. Decide which production method would be most suitable for each of the following products:
   - plastic school chair
   - poster to advertise a school production
   - fizzy drink can
   - unique piece of furniture. (4)

> **KEY TERMS**
>
> **bespoke:** made to the specifications of an individual customer, usually resulting in an original and unique one-off item

1·8·3 Design ideas

> **LEARNING OBJECTIVES**
> By the end of this unit you should:
> ✓ understand different strategies for starting to create ideas.

STARTING POINTS FOR DESIGN IDEAS

When designers start putting ideas onto paper and sketching out their thoughts for solutions to problems, they often need stimuli to help with the creative process. There are many starting points for creating ideas to help designers find inspiration for the **form** and/or **function** of their designs.

Organic shapes, geometry and nature

△ Organic patterns

△ Animal patterns

△ Geometric patterns

△ Geometric natural patterns

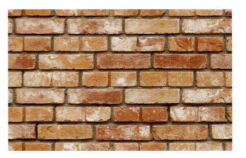
△ Man-made geometric patterns

Cultural inspiration

△ Islamic

△ Celtic

△ African

Biomimicry

By looking closely at structures, shapes and forms of natural things around us, designers have created interesting solutions to problems.

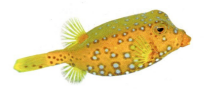

△ Mercedes Benz produced a car called the 'Bionic' which was inspired by aquatic life forms.

△ Toe shoes inspired by the gripping abilities of lizard's feet.

△ Velcro was inspired by plant seeds that stick to other materials.

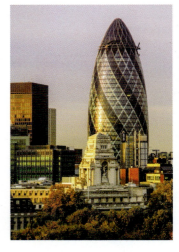

△ The 30 St Mary Axe building in London (widely known as The Gherkin) was inspired by shapes in nature, possibly by a flower bud.

Examining existing products and classic design icons

Looking at existing products and products from the past can provide inspiration. As well as the famous anglepoise desk lamp, have a look at the iconic Coca-Cola bottle and London Underground map.

> **TOP TIP**
>
> Looking at a blank sheet of paper is not usually a great way to start thinking of design ideas. Use inspiration that is all around you.

△ The anglepoise lamp is inspired by the human arm.

Redesigning products with the SCAMPER approach

SCAMPER is an exercise where designers take a design and change it slightly in the following ways:

- **S**ubstitute: Take a shape or form and substitute it with something else.
- **C**ombine: Combine a shape with another.
- **A**dapt: Can a part or component of the product be changed or exchanged?
- **M**odify: Can a shape or component of the product be distorted or changed?

- **P**urpose: Can the part be used for something else?
- **E**liminate: Are there any parts or components that are not necessary? Could they be removed while still maintaining function?
- **R**earrange: Can the design be rearranged while still maintaining function?

It is a good idea to repeat the SCAMPER process three times, to see if your original design has improved, and to inspire your thinking!

Substitute   Combine   Adapt   Modify   Purpose   Eliminate   Rearrange

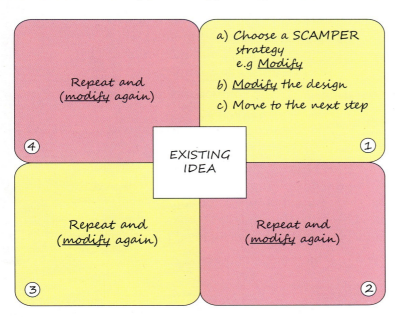

SKILLS ACTIVITY

Select a product in the room you are sitting in and apply the SCAMPER approach to develop it.

KNOWLEDGE CHECK

1. From your understanding of iconic design, list five products, past and present, that you feel fit into this category. Explain your choices. (5)

form: the physical shape and appearance of something

function: the job that a product is designed to do

1·8·4 Identifying constraints

LEARNING OBJECTIVES

By the end of this unit you should:
✓ understand how to plan the stages of manufacture.

PLANNING

The importance of good planning

Once the initial idea, evaluation and further development of a design has been completed, it is time to consider how the end product will be made. Depending on the materials to be used in the production of a product, the planning may take different forms. For example:

- you may be building circuits using electronic components as well as housing for it
- if you were producing a prototype, a model using card and different printing processes could be used, or a mechanical model with levers, linkages and circuits from kits.

For your final product, you will probably require different materials and processes to construct it, and the information needed to do this will have to be clear and accurate. Often this information comes in the form of universally understandable drawings. You will need to:

- produce detailed final drawings, which may include:
  - orthographics
  - isometric or perspective drawings
  - exploded views
  - circuit diagrams
  - designs for moulds
- produce a list of materials and components you require, and work out their costs
- obtain the materials you will need to make your product
- check that you have all of the equipment, machines and processes available to make the product. If there are processes you do not know how to complete or you do not have the skills or machinery/equipment to carry them out, you may have to consider alternatives
- plan the sequence of making, including safety considerations which include risk assessments and quality control checks at each stage.

TOP TIP

Thorough planning will save a great deal of time and frustration in manufacture. Take time to do the planning accurately and in detail.

Using a table to plan

A tabular plan is a logical and sequential way of working through a set of instructions. It can be very thorough and detailed to ensure the instructions are clear and that nothing is missed. For example:

Process	Components	Hand tools, machines and equipment	Quality check	Health and safety	Time allocation	Time taken	Comments

Using diagrams to plan

Using diagrams to plan and give visual instruction can be a very simple way to convey complicated instructions. Note how the examples below both give clear instruction for the intended outcomes.

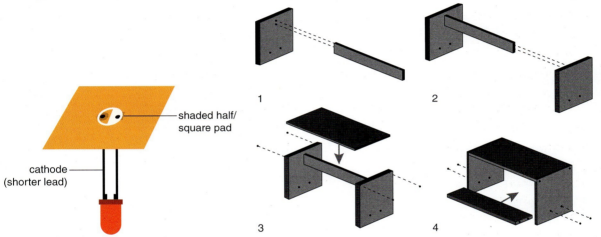

△ A diagram showing how to fit an LED correctly △ A diagram showing how to assemble a piece of flatpack furniture

Flowcharts

A flowchart is a universal drawing method for giving instructions. In a manufacturing and planning context they are useful for listing a set of instructions in a process. They often include decisions and checks at each stage, to ensure the correct sequence is followed.

SKILLS ACTIVITY

Think of a simple chore that you do every day and write a plan or produce an instruction table explaining how to undertake the task.

KNOWLEDGE CHECK

1. What type of diagram would you use to give clear instructions of a component size? (1)
2. What type of drawing would you use to demonstrate the form of a final product? (1)

1·8·5 Evaluating against the specification

LEARNING OBJECTIVES

By the end of this unit you should:
✓ understand how to evaluate your design against its specification.

A DESIGN SPECIFICATION

A design specification is a list of requirements that the completed design must have to be considered a success. The specification should include what the product will do; that is, its primary and secondary functions. Here is a set of considerations your specification may refer to:

- where the product will be used
- target user group
- performance
- ergonomics and anthropometrics
- aesthetics: in the style of …
- safety aspects
- weight/height
- materials
- will it need instructions?
- finish
- testing methods
- packaging
- maintenance
- costs
- storage location
- shelf life
- standards; for example BSI/CE American National Standards Institute (ANSI), International Organization for Standardization (ISO) and Snell.

Remember – a design specification is a working document and will need reviewing and evaluating with the client and user on a regular basis.

EVALUATING IDEAS

Once a designer has a range of ideas, they need to be compared carefully with the specification to make sure the designs meet all the necessary requirements. If not, the ideas will need to be developed.

▽ Idea evaluations

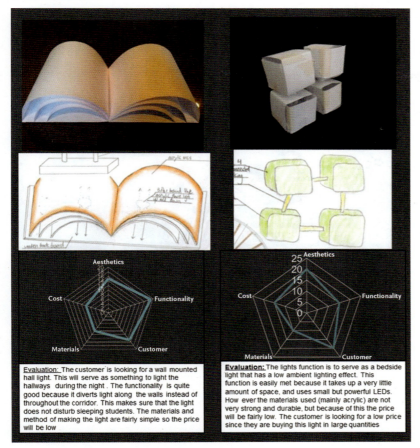

Evaluation against specificaton in a table format

Ideas and final products can be evaluated in different ways. Have a look at the evaluation in unit 1·4. When evaluating, make clear the outcome of each specification point and also explain how the design can be modified to develop the product. Evaluating is a reflective task which allows you to assess your design and progress. Remember: if, after evaluation, you have to redo work, this is positive as it demonstrates the product will improve and be better for the client and users.

TOP TIP
Do not be afraid of constructive critisism or be upset if clients and users are critical of your designs. Use their comments as a basis for development and improvement.

SKILLS ACTIVITY
Write a specification for a pencil case and then create three designs for it. With a colleague, evaluate the designs against the specification.

KNOWLEDGE CHECK
1. Explain why evaluation is an essential part of the design process. If designers did not evaluate, what would be the consequences? (2)

1·8·6 Understanding the relevance of function

LEARNING OBJECTIVES
By the end of this unit you should:
- ✓ understand the terms associated with aesthetics, and relate this to design ideas and the work of others
- ✓ understand the term *function* and how it relates to design.

THE FUNCTION OF A PRODUCT

When a product is designed it often has a number of functions. These are often classified as either primary or secondary.

The **primary function** of a product is usually what it is designed to do. **Secondary functions** often relate to the use of the product through its life, such as cleaning, maintenance, styling and storage. For example, a business card has a primary function of giving information about a person professionally. The secondary function relates to the look of the business card, the type of card, the logo, font and layout. They all represent the company, its aspirations and branding.

△ A business card layout

An electric guitar has a primary function of playing music. Its secondary functions are based on styling, maintenance of the guitar, where to fit the cables and where they are stored, as well as the storage of the guitar itself when it is not in use.

Good design will obviously look at the primary function but will also consider all aspects of any secondary functions.

The two lamps below were made by students in response to the same design brief. Their interpretation is very different but the primary function is the same. They are both desk lights made of similar materials. The difference is in their secondary function-their appearance. This comes down to personal taste and preference.

△ A classic guitar design

△ The primary function of these desk lamps is the same, but the different interpretations of the same brief resulted in their different appearances (secondary functions).

FORM FOLLOWS FUNCTION

'Form follows function' is a quote attributed to the modernist American architect Louis Sullivan, and has been discussed and debated in design and architecture for many years. The basis of the quote is that when designing a product or building, what it is intended to do should be considered first and what it will look like should be considered second. The style and aesthetics of products and buildings relate to personal preferences. These preferences change over time.

These examples of footwear all have the same function. They all do the same thing but for different occasions. They are all footwear, yet they are all very different.

SKILLS ACTIVITY

Choose four products with the same primary function and compare their secondary functions. How do they differ?

KNOWLEDGE CHECK

1. What is the primary function of a product? (1)
2. What is the secondary function of a product? (1)
3. Why do designers need to consider secondary functions? (1)

REFLECTIVE LOG

The famous Philippe Starck Juicy Salif lemon squeezer is a great discussion piece when looking at primary and secondary functions of a product. Look at the image. What do you think about the product's styling? How does it work? Where would you keep such an item? How would you clean it? Discuss your thoughts with a colleague, then with a friend who is not studying design, and compare your responses. How do you think the design could be improved?

KEY TERMS

primary function: the main purpose of what a product is designed to do

secondary function: a function that a product has in addition to its main (primary) function

1·8·7 Aesthetics

LEARNING OBJECTIVES

By the end of this unit you should:
✓ understand the relevance of aesthetics, in terms of line, shape, form, proportion, space, colour and texture, in your design solutions and in the work of others.

WHAT IS AESTHETICS?

Aesthetics is concerned with the visual quality of an item or design. It is what makes an item beautiful. What we consider to be beautiful is different for all of us, so the subject of aesthetics is an interesting one. However, having an understanding of the basic principles is important to help us design products that people will like to buy, and enjoy using.

When designing and when looking at other people's designs, it is important to consider the following:

Shape

A shape is the outline or silhouette of an object. By using different shapes, designs can be made more elegant, slim, fat or chunky. Shapes can be organic, which means they are influenced by natural objects such as plants or rocks. They can also be geometric, which means influenced by shapes like triangles, circles and other polygons.

△ A Verner Panton chair

Form

A form is a three-dimensional solid object. Forms such as prisms, pyramids, spheres and cylinders can be cut to produce new shapes with endless design possibilities.

Proportion

Proportion is related to size: how the height and width of an object relate to each other. If something is out of proportion it does not look correct to some people and is not in harmony. There are scientific/mathematical ways of defining proportion but often it is simply defined by the human eye. The *Vitruvian Man* by Leonardo Da Vinci is a classic example of demonstrating symmetry and proportion.

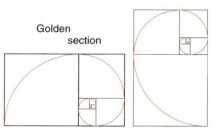

△ Two spirals in proportion

Balance

When positioning parts, images and components on a product, location is important. To look correct they need to be balanced against the other aspects of the design, or they will look out of place.

△ A balanced radio design

Symmetry

In addition to balance, symmetry is important in positioning items in a design. Often asymmetry – the absence of symmetry – is used in design to make things more interesting.

Pattern

Adding pattern or motifs to a design can get rid of plain surfaces and add interest. Patterns can be made of grids of repeating shapes or random lines. They can be organic and/or geometric.

Colour

The use of colour in design is very important for many reasons. Colour can be used for function as well as aesthetics; for example, using green for start and red for stop. It can also be used to appeal to a particular customer or market: for example, in European culture, gender-based products for babies are often either pink (for girls) or light blue (for boys). Using warm and cool colours, bold colours, tints, shades and natural colours can have dramatic effects on styling in design ideas.

Texture

Texture can be added to products for functional reasons such as grip, but can also be for decorative reasons. Some materials such as wood and leather have natural textures, but other materials have textures added during the manufacturing processes, such as plastic chairs in a classroom or the handle of a golf club.

△ The yin and yang symbol uses elements of symmetry.

△ A selection of patterns that can be used to make products more interesting.

△ Using colour for elevator buttons is useful for indicating their function.

△ The texture on a golf club handle is mainly functional because it is used for grip.

> **TOP TIP**
>
> The Fibonacci series of numbers and the Golden Ratio (also referred to as the Golden Mean) are two mathematical formulas that are often used in design. Use the internet or library to research them, and use this as a starting point for your design ideas and development.

SKILLS ACTIVITY

Look at the work of the modern designers Karim Rashid, Dieter Rams, Marc Newson, James Dyson, Sori Yanagi and Philippe Starck. What differences can you see in the styles of their work? How do they use the principles mentioned in this unit in their work?

KNOWLEDGE CHECK

1. In your own words, define 'style'. (2)
2. Explain why balance and symmetry are important in design. (2)
3. In what ways can colour be used in design to change products? (2)

1·8·8 Style and design movements

LEARNING OBJECTIVES
By the end of this unit you should:
- ✓ understand the basics of style
- ✓ understand the principle design movements.

STYLE

When undertaking a new design project, it is useful to look at existing designs to get an idea of what has already been done in terms of styling. This is often a good starting point.

Style is the interaction of aesthetic elements such as shape, form, pattern, texture, colour and proportion. Styles are dependent on fashion, people and locations, politics and materials, and so they are constantly changing and developing.

△ A chair in a classical style

Styles can loosely be categorised as:

- classical
- modern
- functional
- hi-tech.

Of course, within each of these groups there is a huge range of different styles. The products in the photographs opposite are both chairs but they have different styles. Which ones you find attractive, or not, will vary according to your tastes. Designers need to consider styles and visual elements to fit the intended environment and end users.

△ This Barcelona chair is a modern style

DESIGN HISTORY

The work of past designers and architects is interesting and a useful starting point when designing new products. Over the past 2000 or more years, technology, science, materials and making have changed considerably. Over time, the buildings we live in and the products we use have changed. The major changes have taken place in the past 250 years since the industrial revolution. With industrialisation came mass production through mechanisation. Before mechanisation, products were hand crafted with great skill.

The main design style movements of the past 150 years can be loosely categorised. Their styles are a mixture of shape, form, pattern, colour and texture, and have been influenced by social and political situations. A number of these styles are listed on the next page. Looking at them closely could help you with your own design ideas.

Arts and Crafts

Possibly considered the first design movement; Arts and Crafts was a movement supported by a group of people who developed a philosophy of craftmanship over mechanisation. They were reacting against industrialisation, the machine age and mass production. It was started by William Morris in the 1880s. They where influenced by nostalgia and gothic architecture, and their designs where very floral with animal and Celtic motifs. Arts and crafts followers and members where highly skilled in their trades.

△ An example of a William Morris pattern

Art Nouveau

Art Nouveau developed the Arts and Crafts movement in the sense that their styling was very organic, using natural forms and curved lines. However, unlike the Arts and Crafts group, Art Nouveau was accepting of mass production techniques. It was mainly a European and North American based movement with many variations. Some of the main designers where Charles Rennie Mackintosh and Louis Comfort Tiffany.

△ A Tiffany lamp

Art Deco

Art Deco was a reaction to the organic motifs of Art Nouveau. Art Deco was influenced by Cubism, which is very geometric in styling with strong vertical lines, zig zags and sharp angles. Colours were often bold and materials were new and modern. The main designers were Eileen Gray and Clarice Cliff.

△ A radio deisgned in an Art Deco style.

Bauhaus

The Bauhaus was a school of art and architecture founded in Germany by Walter Gropius. It was pioneering in combining art and technology. The school designed prototypes, which were intended for mass production. The styling was strongly simplified, while still maintaining function. Key designers where Marcel Breuer and Wassily Kandinsky.

△ The Bauhaus school in Germany

Modernism

Modernism covers a very wide time period and a number of movements, but in general it tried to simplify and minimise design, making items standardised and cultureless. It uses new and modern materials and manufacturing processes. Key designers are Isamu Noguchi, Charles and Ray Eames and Verner Panton.

△ A Noguchi table

Postmodernism – Memphis

Postmodernism was a reaction to the minimalism period. It used bold, shocking and strong primary colours with deep texture in its products. Geometric shapes with plastics and laminates were extensively used. A key designer was Ettore Sottsass.

Blobism

Blobism is a very modern style with strong curvature that is often irregular. The products are mass produced and this is enabled by heavy use of CAD/CAM. Blobism frequently uses bright colours, and style is minimal and often contrasting. Materials are usually manufactured and not natural such as plastics, composites, alloys and metals. Key designers include Karim Rashid, Jonathan Ive and Marc Newson.

NOW AND NEXT ...

What design movement are we in now and what is next? Some would say we are in the maker movement: a fast-growing group of technologists, designers and creators who use new technology to develop Do-It-Yourself (DIY) products. This movement encourages creativity and prototyping but does not have any particular styling approach.

△ An example of postmodern architecture

TOP TIP

Over time, you will create your own style when designing. However, look at the work of past designers and styles to give you initial ideas and inspiration.

DESIGN IN ACTION

The company Apple has been extremely successful because of their product design. Jonathan Ive is their chief designer and responsible for much of the Apple product range. He has been greatly influenced by the work of Dieter Rams who was the chief designer at Braun, Germany in the 1960.

SKILLS ACTIVITY

Look around your home or home town. Make a list of items that might fit into each of the design styles mentioned in this unit. This may include buildings.

KNOWLEDGE CHECK

1. Explain why it is important to look at the work of past designers in your own creative work. (4)
2. When we talk about style, what are we considering when it comes to products? (2)
3. Look at the style and design movements again and choose one of the products photographed. From the information given, explain why you think it fits into that design movement. (4)

1·8·9 Using models to test proposals

LEARNING OBJECTIVES
By the end of this unit you should:
✓ understand the importance of anthropometrics and ergonomics in design
✓ be able to use models to test proposals.

USING SCALED ERGONOMES

Scaled models made from a wide variety of compliant or soft materials can be useful in testing designs against human sizes and anthropometric data. A scaled model could take the form of a 2D drawing or a 3D model.

Scaled, full-size drawings can be used to check sizes against potential users.

Ergonomes (a 2D human shape) and manikins (3D human form) can be used to see the relationship between the product and the user.

△ An ergonome next to a shopping trolley

Two dimensional ergonomes can be made out of hardboard. They are made by first researching the sizes of a particular user group. The average data (compared against scientific data gained from books and the internet) is then converted into body parts. They are useful for testing design prototypes and comparing potential users against scaled drawings.

FROM DESIGN TO MODELS TO TESTING

The images below relate to an exercise where the students are to design scissors for a specific user group. The images show initial ideas followed by models made from a variety of materials. Then, a model of a human hand is used to check and test the sizes. This testing is very important at the early design stage to check the function of the product and its relationship with the user. The next stage would be to get a specific user to test the models.

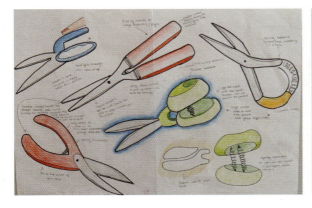
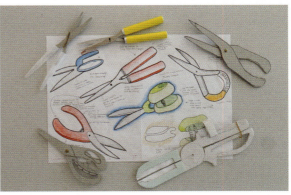

 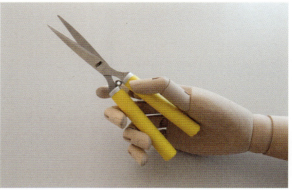

△ A modelling exercise for designing scissors

TOP TIP
When designing, always keep the users' needs in mind and work with them closely to ensure the product meets their requirements.

INCLUSIVE AND EXCLUSIVE DESIGN

Inclusive design is where everyday products and services are designed to be accessible to, and usable by, as many people as possible. One example is a magazine.

Exclusive design is where products and services are designed for a specific user group. This could include bespoke manufacture of elite products, but in this context refers to design products and services for user groups with particular needs. This may include people with physical or mental disabilities, and patients who are recovering from accidents and trauma. It may also include user groups such as the elderly or children. For example, an audio or braille version of a magazine.

When designing products, it is important to identify whether the end product should be inclusive, with a broad range of intended users, or, if it is exclusive, intended for a niche market. When modelling for an exclusive group, wherever possible make sure real users test your designs and provide feedback.

SKILLS ACTIVITY

Take a plain piece of paper and draw around your hand. Measure the lengths of your fingers, width of your palm and other dimensions. Consider how a designer would use this information when designing a handheld product such as a mobile phone.

KNOWLEDGE CHECK
1. How does modelling a design help its development? (2)
2. What is an ergonome and how can it help you when designing? (2)

1·8·10 Modelling

> **LEARNING OBJECTIVES**
> By the end of this unit you should:
> ✓ understand the need to model and test proposals.

THE IMPORTANCE OF MODELLING AN IDEA

Modelling is an important part of the development process. Making a physical model that can be touched and manipulated is the only method that will enable you to interact with the product, testing how it feels and also experiencing the ergonomics of the shape, form and size. Modelling can take a number of forms, including:

- test modelling – using computers to simulate a design and specific conditions
- scale modelling/prototyping – using handmade, scale models that work and function like the intended design
- 3D CAD modelling – 3D simulations with renderings to get a lifelike representation of a design.

Handmade models

Producing handmade models with kits, a range of materials from paper to clay, and construction models such as circuit boards, are the main ways of modelling. They can be very easy to produce with basic equipment and provide immediate results.

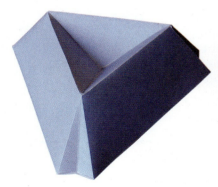

△ An example of a paper model

△ A student cutting out a balsa wood model

> **TOP TIP**
> Use recycled cardboard and basic tools, such as a cutting matt, craft knife and glue gun, to create a range of quick sketch models to demonstrate an idea, test a concept or develop a design. This process is quick, and iterations of a design can be achieved very easily.

CAD and CAM modelling

Computer modelling through CAD can be used to create both 2D and 3D models, which can be tested through simulation software. Electronic circuits can also be designed and tested in this way. The benefits of CAD modelling are:

- They can be changed easily.
- They can be sent to a 3D printer or CNC machine anywhere in the world.
- Multiple copies can be made.
- Complex organic shapes can be created and modelled.

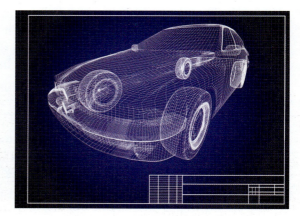

△ Examples of CAD blueprints

Models can also be made through CAM – rapid prototyping and 3D printing. For example, a structural engineer may design a tall building and expose it to conditions like wind and earthquakes to test its design.

Modelling with CAD and CAM allows the designer to change different components without the expense of actually having to purchase a range of items, reducing the cost of development and faults at a later date.

△ A rapid prototype of a building rendered in CAD

SKILLS ACTIVITY

Think of designing in industry and the commercial world. Make a list of examples where models may be used in the design process to assist in the creation of a design.

KNOWLEDGE CHECK

1. Give five benefits of modelling designs. (5)
2. Name five materials that could be used for modelling designs. (5)

1·9 Environment and sustainability

1·9·1 Forms of energy

> **LEARNING OBJECTIVES**
> By the end of this unit you should:
> ✓ know about a range of conventional energy forms
> ✓ know about a range of alternative energy forms
> ✓ understand the difference between finite and infinite sources of energy.

CONVENTIONAL ENERGY SOURCES

The industrial and commercial world we live in relies heavily on energy (electricity) to drive economies and to power our increasingly complex lives. Many of the products we take for granted in our homes, schools, hospitals and transport systems require large amounts of energy.

The main sources of energy in many countries come from fossil fuels such as oil, gas and coal. These three fuels, as the name 'fossil fuels' suggests, are created from animals and plants that died millions of years ago and over time have been covered by layer upon layer of natural deposits which have slowly compressed them. The addition of heat and lots of pressure have converted them into carbon-rich deposits that are extracted from underground and converted into energy.

Fossil fuels

Fossil fuels are used extensively in modern industrial countries such as North America, India, China, Australia, and in Western Europe. Developing nations in Africa, Southern America and Asia are increasingly using more and more fossil fuels.

- **Coal**
  Coal is very common and found all over the world. It is relatively easy to extract but extremely labour intensive and dangerous. Coal has been mined for hundreds of years. When burnt, it produces high heat which is used for cooking, heating and also for creating electricity in power stations.

- **Oil**
  Oil is found deep underground in large reservoirs. Holes are drilled into the ground until the oil is found. It is then pumped out of the reservoir and transported. Oil can be converted into fuels such as petrol and diesel and also, through different processes, turned into the many plastics we have all around us. Oil is found in many countries and is an extremely valuable resource.

- **Gas**
  Gases are found underground, usually near oil fields and also in gaps in rocks. Natural gas is extracted from underground. Since it is very explosive, transportation over along distance is dangerous and it is therefore often used near to the place of extraction. However, it can be converted to a liquid for transportation and then used as a heating fuel, and also to create electricity in power plants.

All three fuels take millions of years to produce and therefore can only be extracted once. They are finite: in time, they will run out. When they will run out is debatable but, with an increasing world population, increasing personal consumption, and an increasing reliance on power, this time may come quicker than we think.

Nuclear fuel

In many developed nations, nuclear energy is used. While controversial because of the long-term effects of the waste material produced by the nuclear process, it is considered relatively clean compared to burning coal, for example. However, it is very expensive to set up.

RENEWABLE ENERGY SOURCES

Renewable energy sources are alternatives to fossil fuels and nuclear energy. They are considered infinite as the sources of the energy are produced naturally by the Earth. The sun, wind, seas, rivers and underground sources of heat can be utilised to produce energy. While at present renewables are used in small quantities compared to fossil fuel and nuclear energy production, this will increase over time when fossil fuel supplies start to run out.

Different renewable energy forms are:

- **Solar photovoltaic:** The sun's rays are converted to electricity from solar panels.
- **Wind:** Wind power is used to turn blades on a turbine and is then converted to electricity.
- **Hydroelectricity:** Water from a resevoir naturally flows down-river through turbines which generate electricity.
- **Solar thermal:** The sun's heat is used to heat water, for example, for showers and washing.
- **Geothermal:** Cold water is pumped deep into the ground and returns hot for heating purposes.
- **Wave energy:** The power and movement of waves in the sea are converted into electricity.
- **Tidal energy:** The power of the daily tide in a river estuary is used to drive turbines, which then generate electricity.
- **Biomass:** When living organisms (such as plants) die, the gases produced are burnt to generate electricity.

△ A photovoltaic solar panel

△ A wind generator farm

△ A hydroelectric dam

SKILLS ACTIVITY

For the table of countries below, which type of energy sources do you think are used most and why?

Nepal	Brazil	Denmark
Singapore	Iceland	China

EFFICIENT AND EFFECTIVE USE OF ENERGY SOURCES

The energy that many of us take for granted every day is usually made from fossil fuels. As mentioned earlier, these are finite. While we could transfer to alternative, more ecological sources these types of energy can sometimes be expensive. A simple way to ensure fossil fuels last longer is to manage the energy used at home, school and in commerce and industry better. Switch off lights when they are not needed. Close the doors so valuable hot or cold air is kept indoors to keep people comfortable. Do not take the car when you could walk or cycle – the exercise is good for you, too! The 'One Laptop Per Child' initiative in schools took away the need for charging by having a winder to provide energy. The innovator Trevor Baylis designed the clockwork radio that could be used anywhere in the world. It also did not need batteries but used mechanical energy and a solar panel for power.

△ Schoolchildren using laptops from the One Laptop per Child scheme

HOW DESIGNERS CAN REDUCE ENERGY CONSUMPTION

Good designers are well informed, environmentally responsible and consider many things when they are going through the process of creating new products. In terms of energy consumption, the designer should consider the following:

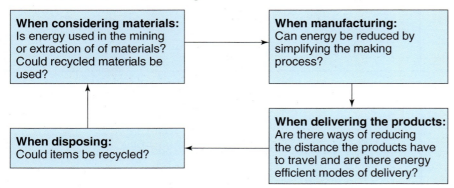

THE SIX 'R'S

Many people are familiar with the environmental three 'R's: Reduce, Reuse and Recycle. Many people now consider an additional three to make it six:

- **reduce** your use of electricity, quantities of products, and so on
- **reuse** packaging, materials and items a second time – upcycling
- **recycle** materials and put them to use in other applications
- **rethink** the way you do things and change the way you think and work
- **repair** items rather than throwing them away and buying new ones. Design products that are easily repaired
- **refuse** to buy a product if you do not need it.

RECYCLING

Recycling, similar to reusing, reduces environmental damage by using products again, which eliminates the need for more raw materials.

Products such as aluminium cans, some glass drinks bottles, clothing, paper and card are commonly recycled. However, we are still consuming more than we recycle and generate far more waste than before. More and more people and companies these days regularly recycle, but more still needs to be done. That is why the 6'R's are important and why designers are in a great position to create products that are more sustainable.

△ Packaging recycling symbols

△ Aluminium recycling symbol

△ Glass recycling symbol

Food packaging, for example, accounts for much of our domestic waste. Most food packaging derives from plastic, which comes from the finite resource of oil. Switching to more sustainable packaging would be very helpful. Recent developments in bioplastics have created products that will naturally degrade when they are thrown away.

Many of us have a number of mobile devices such as phones, tablets, laptops, and so on. The batteries in these products are extremely dangerous when thrown away. They are manufactured with poisonous chemicals that can leak into the water supply that we drink. Rechargable batteries help but are still an issue. What is the solution?

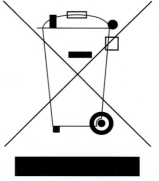

△ Waste electrical and electronic equipment recycling symbol

SKILLS ACTIVITY

Discuss in pairs or in groups possible solutions to the issue of throwing away batteries. What could be done with them that is more environmentally friendly?

DISASSEMBLY

Many products are designed so that once they are assembled they are difficult to take apart and maintain. This means that people throw them away without trying to fix them. These products usually end up in landfill sites. An alternative approach is to design products that can be taken apart using common tools and equipment. This would make it easier to replace or repair parts and modules when they break. This extends the life of the product and reduces the need to use more resources and raw materials, reducing waste.

△ An exploded view of a mobile phone. Some of its parts could be recycled.

Some manufacturers do not want this to become the norm as their aim is to sell more products when older products break, so they can make more sales and more profit. This leads to the use of more raw materials and is not good for the environment. In addition, it could also be seen as unethical. This approach is very short-sighted.

Another feature of designing for disassembly is where a product is designed to come apart easily at the end of its life. This allows for some of the parts or subassemblies to be easily accessed so they can be recycled.

PLANNED OBSOLESCENCE

Planned obsolescence is the practice of designing products with a limited life. Manufacturers intentionally design the product so that it becomes obsolete after a certain period of time. This obsolescence may come in many forms, including:

- software upgrades
- material failure
- component failure
- product assembly problems
- compatability issues.

Take, for example, a desktop printer. The product will fail after a period of time but will last long enough for the customer to be satisfied with the product and company. The customer would usually have had good use of the machine and will generally purchase another product from the same company, as they have developed a sense of brand loyalty.

Some printers are designed so that after printing a certain number of sheets of paper, some weaker components may break. The manufacturers work out that this would be after one year or so, after the warranty for the product has expired. The customer has a choice to either replace the part or buy a new product. Replacing parts is sometimes very expensive and specialist tools and engineers are needed. The cost of doing this is often much the same as buying a new product. So, does the customer replace the printer? The manufacturer hopes the customer has liked their product and will replace it with the newest version.

However, planned obsolescence can be a positive concept. Food manufacturers have sell-by dates so products are eaten before they go off and become a health hazard. This is for consumer safety. Some products made of Polylactide (PLA), such as plastic cutlery and plastic bags, biodegrade quickly so need to be used and thrown away within a limited time period.

MATERIALS RECYCLING

Many materials can be recycled or reused with the correct processing procedures. For example, most plastics can be recycled. Plastics can be categorised into two main groups: thermosetting and thermoplastics. The latter can be recycled. Many products that are made of thermoplastics are stamped with a recycling triangle containing a number. The number indicates the type of plastic the product is made from, so these plastics can be grouped together for recycling.

△ The system of plastic recycling symbols

Some materials are joined together when they are manufactured. These are called composite materials. Composites are usually very difficult to recyle back into their original materials. They include materials such as concrete, drinks cartons, MDF board, and printed circuit boards (PCBs). See units 3.3 and 3.5 for more information.

These symbols below are becoming more widely used on products and packaging to encourage consumers to think about recycling and the environmental impact of waste.

△ Recyclable steel symbol

△ Compostable symbol

△ The green dot symbol is used when a manufacturer has contributed to the cost of recycling their products

SKILLS ACTIVITY

Make a list of common materials that can be recycled and ones that cannot.

Materials that *can* be recycled	Materials that *cannot* be recycled

DESIGN IN ACTION

The organisation Practical Action (practicalaction.org) is a charity that supports poor men, women and families across the world and uses sustainability as the foundation for their charity work. Their aim is to enable poor communities to build on their skills and knowledge in order to produce sustainable and practical solutions to the problems they are faced with.

KNOWLEDGE CHECK

1. Name four sources of renewable energy. (2)
2. Name ways in which we can make more efficient use of energy to reduce our consumption. (4)
3. Which three of the 6'R's are most recent? (3)
4. What are the advantages of making products that can be disassembled? (6)
5. What is meant by 'planned obsolescence'? (2)

Graphic design has shown remarkable persistence in being able to evolve with the times. As technology has changed, graphic design has changed with it. From the early days of cave painting to the modern information age of touchscreens, brands and advertising, the need for humans to communicate quickly and effectively has remained. Graphic design is about visual communication and, in our modern world, where access to quick information is essential, it plays an increasingly vital role in meeting this demand.

Graphic design takes many forms and is ubiquitous in its presence around us. Logo design helps us recognise trusted brands; packaging allows us to identify products we want; posters and leaflets communicate ideas, concepts and events; and we read newspapers and magazines for up-to-date information. When we turn on the television, our favourite programmes are preceded by title sequences designed by a graphic designer, and our favourite movies are full of special effects. On the internet we are able to navigate websites easily to find the information we need. Our smartphones contain apps that allow us to read the news, interact with other people and play games. All these products we use are made well because of the drawings supplied by the designer to the manufacturer.

The Graphic Products section of this course will teach you how to communicate your ideas effectively. You will develop a number of techniques: from improving your formal drawing skills and understanding the importance of modelling prototypes, to using computer-aided design and ICT to develop and present ideas quickly and easily. You will also learn about how graphic products are made in industry. In an age of mass media and mass production, industrial printing methods and computer-aided manufacture are essential in meeting a global appetite for information, services and products.

STARTING POINTS

- What type of drawings are best to communicate different ideas?
- How can you draw basic shapes and enlarge and reduce them accurately?
- What are developments and what is their importance in graphic products?
- What tools are available to help communicate ideas effectively?
- How can you lay out and present ideas in an attractive way that enhances communication?
- How is ICT used to increase accuracy and productivity?
- How are graphic products printed and made in volume?
- What materials are available to allow effective communication through modelling?

SECTION CONTENTS

- **2·1** Formal drawing techniques
- **2·2** Sectional views, exploded drawings and assembly drawings
- **2·3** Freehand drawing
- **2·4** Drawing basic shapes
- **2·5** Developments
- **2·6** Enlarging and reducing
- **2·7** Instruments and drafting aids
- **2·8** Layout and planning
- **2·9** Presentation
- **2·10** Data graphics
- **2·11** Reprographics
- **2·12** Materials and modelling
- **2·13** ICT
- **2·14** Manufacture of graphic products

2·1 Formal drawing techniques

ORTHOGRAPHIC, ISOMETRIC, PLANOMETRIC AND PERSPECTIVE

LEARNING OBJECTIVES

By the end of this section you should:
- ✓ be familiar with a range of formal drawing techniques including first- and third-angle orthographic projection, isometric, planometric and two-point perspective
- ✓ understand how to use British Standards correctly.

ORTHOGRAPHIC DRAWING

In design, it is important that formal drawing techniques are used to convey important information about the design to either the client or the manufacturer. **Orthographic projection** is a form of drawing which shows three sides of the product and is drawn to scale. An **orthographic drawing** is laid out in either **third-angle projection** or **first-angle projection**. Two different layouts of a sign are shown below.

Third-angle projection shows the plan view of the sign in the top left above the front elevation. First-angle projection shows the plan view in the bottom right below the front elevation.

Notice the dotted lines on the plan view and the front elevation. This type of line is used to show **hidden detail**. These are parts of the drawing that cannot be seen if you were looking at the object, yet still need to be shown in the drawing.

Third-angle projection is a cone with the plan view on the left. The crosshairs and line show the centre point of the cone.

First-angle projection is a cone with the plan view on the right. The crosshairs and line show the centre point of the cone.

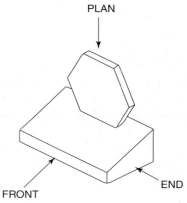
△ A pictorial view of a product

THIRD ANGLE PROJECTION

FIRST ANGLE PROJECTION

△ Projection symbols

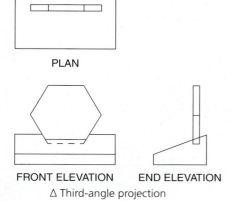
△ Third-angle projection

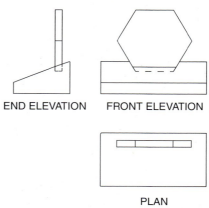
△ First-angle projection

SKILLS ACTIVITY

Find an object in your bag or pencil case, such as a pencil sharpener. Keeping one eye closed, look at it from the top and sketch what you see. Now turn it to the front and sketch what you see. Do the same from the side. Make sure you only draw the one side you are looking at. These three views would form the plan, front and end elevations of an orthographic drawing.

Scale and dimensions

Sometimes the object you are drawing is too big to draw at actual size on paper. Sometimes it is too small. In cases like this you will need to consider **scale**. If your drawing is half the size of the object, the scale will be 1:2 (you can visualise replacing the : with a /). If the drawing is one third the size of the object, the scale would be 1:3. If the object is small and you want the drawing twice as big, the scale would be 2:1. You should show the scale you have used at the bottom of each page. **Dimensions** are measurements of parts of the drawing. Dimensions are usually written in millimetres (mm).

Standards in orthographic projection

Since many products are made in different parts of the world, it is important that a common form of drawing and adding dimensions is used so that the drawings can be understood by designers and manufacturers. There are several different standards such as **British Standards** (BS), International Organization for Standardization (ISO) and American National Standards Institute (ANSI). In your drawings it will be expected that you use British Standards as shown below.

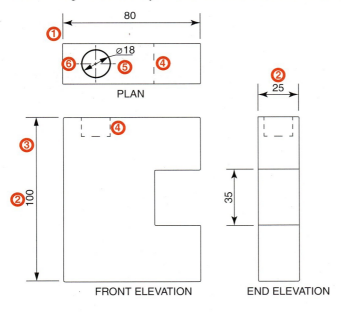

① Projection lines should be drawn off the edges that are being dimensioned but not touching them. They should be half the thickness of the lines of the drawing.

② Dimensions are added above the dimension line if the dimension line is horizontal. If the dimension line is vertical the dimensions are placed to the left and rotated. Units are assumed to be in mm so don't add these.

③ Dimension lines should be half the thickness of the drawing and arrow heads solid black.

④ Hidden Detail are parts of the drawing that are there but can't be seen if you are looking at the object from the view given. They are shown as a dashed line.

⑤ Diameters of circles and curves are shown with a Ø symbol and can be drawn as in the examples.

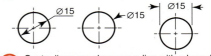

⑥ Centre lines are shown as a line with a long dash followed by a short dash.

△ British Standards are one of the conventions used for working drawings.

SKILLS ACTIVITY

Here a trophy is shown. The trophy consists of two parts: a circular base and a shape, which slots into the base. The assembled trophy is shown to the left. Draw the assembled trophy in third-angle projection at a scale of 1:2.

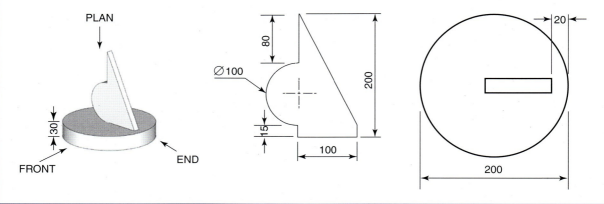

ISOMETRIC AND PLANOMETRIC DRAWINGS

Isometric drawings are three-dimensional drawings at 60° and 30°. Isometric drawings are usually used to visualise products in three dimensions.

Planometric drawings are commonly used for interior rooms of buildings as they can be drawn from the plan view of a room. Planometric drawings are drawn at 45° and 45°.

You can use isometric paper underlay or a set square to help you draw in either isometric or planometric.

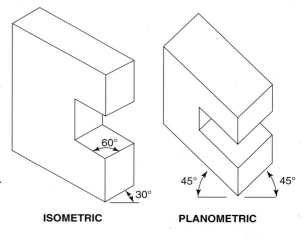

TOP TIP

Whatever the object you are drawing in isometric, it is important you draw a constructional cube or cuboid first, which is the total length, width and height of the object you are drawing. You can then draw the object inside this cube. This is called **crating**. This will help you get the sizes and proportions right when you draw the object.

Drawing circles and arcs in isometric

1. Draw an isometric cube with all sides even.

2. Split one of the cube faces vertically and horizontally down the middle. Then draw a line from the top left corner to the bottom right.

3. Draw two lines from the top middle to the bottom corner and from the top right to the bottom middle.

4. Place the point of your compass on *x* and draw from 1 to 2. You have now drawn two arcs isometrically.

5. Place the point of your compass on *x* and draw from 2 to 3. Finally, place the compass on *y* and draw from 1 to 4.

6. Remove any previous construction lines. If you want to draw a circle on the top plane or the right plane then repeat the steps on these planes.

TWO-POINT PERSPECTIVE DRAWING

Two-point perspective drawing is a type of drawing that gives a realistic representation of a building, interior or product. To construct a perspective drawing you will need a horizon line and two vanishing points (VP). The width and the depth of the object converge towards these two points.

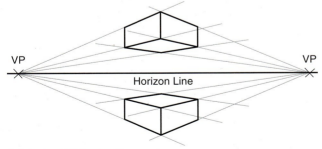

△ A two-point perspective

Notice that by drawing the object above or below the horizon line it, makes it look like you are looking at the object from above or below.

KNOWLEDGE CHECK

1. What tools can be used to draw isometric and planometric? (1)
2. Draw the symbols for first- and third-angle orthographic projection. (2)
3. Explain what is meant by 'hidden detail'. (4)

KEY TERMS

British Standards: the commonly accepted way in the UK of drawing lines and adding dimensions

crating: the practice of drawing inside a cuboid to help the drawing of isometric or planometric projections

dimensions: the measurements of the size of an object

first-angle projection: a method of laying out an orthographic drawing with the end elevation in the top left and the plan view below the front view

hidden detail: part or parts of an object that cannot be seen, shown by a dotted line on the drawing

isometric: describes a method of drawing a three-dimensional object at 60° and 30° angles

orthographic drawing: a method of drawing a three-dimensional object in two dimensions, usually showing a plan view, front elevation and end elevation

orthographic projection: a two-dimensional drawing that represents a three-dimensional object, showing three sides of the product and drawn to scale

planometric: a method of drawing a three-dimensionsal object at 45° and 45° angles

scale: a ratio indicating the size of the drawing compared to the actual object being drawn

third-angle projection: a method of laying out an orthographic drawing with the end elevation in the bottom right and the plan view and front elevation to the left

two-point perspective: a method of making a realistic three-dimensional drawing using two vanishing points and a horizon line

2·2 Sectional views, exploded drawings and assembly drawings

LEARNING OBJECTIVES

By the end of this unit you should:
✓ be able to construct different types of sectional views
✓ be able to construct exploded and assembly drawings.

SECTIONAL VIEWS

Sometimes it is important to show the interior of an object. In this case a sectional view would give the appropriate information to a manufacturer. A **sectional view** or cross-sectional view is a view of an object as if a part of it has been cut away with a saw. The area that 'cuts' is called the **cutting plane**. Below are three different views of an object. The object has a hole through the curved section and a peg inserted into the base on the right side. A sectional view needs to be produced to see the peg inside the object.

△ The object with a hole through the middle

△ The object shown as a see-through view

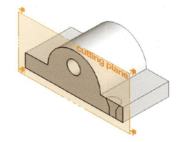

△ The object showing the cut view

On the next page, drawing A shows the plan of the object with a line through the middle (A–A) indicating where the cutting plane is. Drawing B is a sectional view. In a sectional view the part of the object that has been cut is cross hatched at 45°. There is no hatching where the hole is as it is empty space. Notice the hatching in the opposite direction on the peg since it is a separate, solid object.

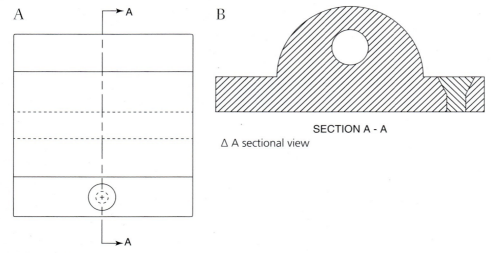

△ A plan of the object and cutting plane A-A

△ A sectional view

PART, REVOLVED AND REMOVED SECTIONS

Below is a part sectional view where a quarter of the object has been removed. In the unsectioned area, hidden detail is not shown.

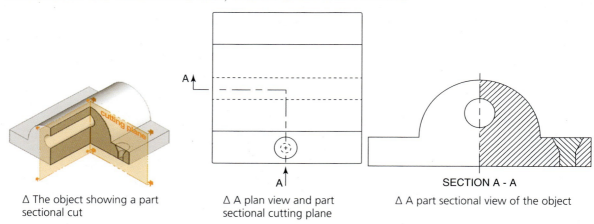

△ The object showing a part sectional cut

△ A plan view and part sectional cutting plane

△ A part sectional view of the object

Sometimes the sectioned part is shown separately from the main drawing. This can help give a clearer view of the sectioned area and allow for dimensions to be added easily. In this example the rivet is shown as a removed sectional view.

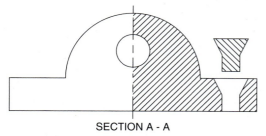

SECTION A - A

△ The object with removed section

A revolved section is drawn on the view as opposed to a separate drawing and shows the section rotated 90°. This is often used to show sections of long, elongated shapes.

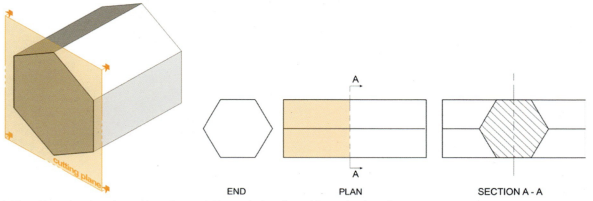

△ The object showing the cutting plane   △ The end elevation with revolved section

SKILLS ACTIVITY

Here is a pictorial drawing of an object. Draw the object as a sectional view along A–A, using the scale 1:1.

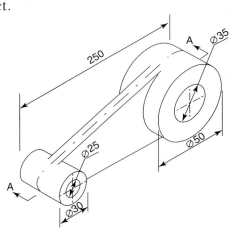

EXPLODED AND ASSEMBLY DRAWINGS

An **exploded drawing** shows different components separated from each other by a certain distance. The parts appear to be floating in space. An **assembly drawing** is the opposite, and shows all the parts assembled. Exploded drawings are useful for showing how pieces fit together and in which order. They can often be found as instructions for assembling flat-packed furniture.

Exploded drawings should be drawn as line drawings in isometric. The components should line up with each other correctly so you need to draw construction lines before completing an exploded drawing.

The object below has three pegs and an additional plug for the hole. The first image is the object with all the parts assembled. The second image shows all the parts separated.

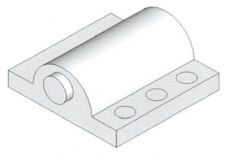

△ A pictorial view of an assembled product.

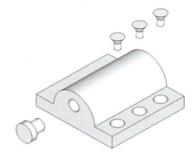

△ An exploded drawing showing the parts separately.

TOP TIP

To construct an accurate exploded drawing you need to use crating as constructional lines in the first image. Make sure these are at the correct angles and in line with the holes or slots on the drawing.

Image A shows the object drawn using crating. Image B shows the completed exploded drawing.

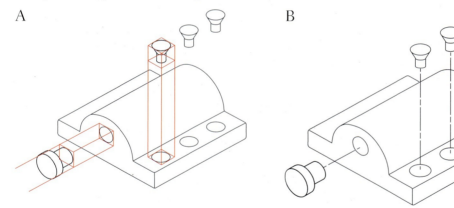

Parts list

A parts list often accompanies an exploded drawing. This shows all the parts and components needed to assemble the product. Compiling a parts list can help keep track of which components are needed for your product and are often used for calculating cost.

Item	Quantity	Unit	Price/unit ($)	Total cost ($)
Red acrylic	1 sheet	3mm sheet	6.63	6.63
Blue acrylic	1 sheet	3mm sheet	6.63	6.63
Coated card	10 sheets	3mm sheet	1.50	15
ABS	200g	1kg spool	20	4
Plastic rivets	10	100pc	0.1	10
			Total	42.26

SKILLS ACTIVITY

Below is a pictorial drawing of a children's toy. The three shapes can be removed from the holes in the base. Also shown is the side elevation of the base of the toy. Draw an isometric exploded view of the toy with the shapes removed. Estimate all sizes.

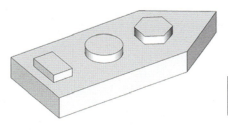

KNOWLEDGE CHECK

1. Explain the difference between sectional, part sectional, revolved sectional and removed sectional drawings. (8)
2. What is meant by a cutting plane? (1)
3. Explain a context in which an exploded drawing may be used. (2)

> **KEY TERMS**
>
> **assembly drawing:** an isometric view showing parts and components assembled
>
> **cutting plane:** an imaginary plane that cuts an object
>
> **exploded drawing:** an isometric view showing parts and components separated
>
> **sectional view:** a view of a face of an object as if it has been cut through. Different types of sectional views include full, part, revolved and removed sectional views.

2·3 Freehand drawing

LEARNING OBJECTIVES

By the end of this section you should:
✓ be able to sketch in a variety of ways using freehand drawing.

Freehand drawing is sketching without the use of drawing aids or grids. Designers often produce freehand sketches and notes to communicate their ideas from written data, visual data and data from tables. It is important to select the most relevant method of drawing to represent the information you have been given.

ISOMETRIC

Being able to sketch in isometric can help you quickly visualise your ideas. Designers often use this type of sketch to quickly get their ideas down on paper, and they can also help convey ideas to their clients since they can give a relatively accurate reflection of what the final design will look like.

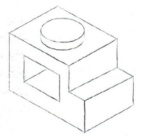

△ A freehand isometric

PERSPECTIVE

Freehand perspective gives more realism to a sketch. You will need to draw a horizon line and vanishing points first. Estimated two-point perspective is often used by architects to show the exterior of buildings. It is more suited to larger scenes such as environments and buildings, rather than smaller objects. Estimated one-point perspective is best suited to drawing rooms and interiors of buildings.

A one-point perspective drawing uses a horizon line with one vanishing point (VP). Lines projected from these vanishing points form the width of the object.

A two-point perspective drawing uses a horizon line with two VPs. Lines projected from these vanishing points form the width and length of the object. See the paragraph on two-point perspective drawing in unit 2·1.

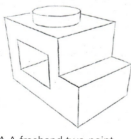

△ A freehand two-point perspective

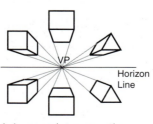

△ A one-point perspective

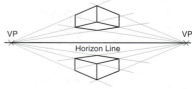

△ A two-point perspective

EXPLODED VIEW

A freehand exploded drawing will allow you to show how pieces fit together. This is a useful type of drawing if your design has many components as it helps to explain how and where parts go together.

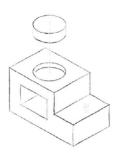

▷ A freehand exploded drawing

TOP TIP

Practice makes perfect. Getting your angles right in isometric freehand drawing can be difficult. You can use a grid underlay to help you practise and finetune your freehand skills, but remember not to present drawings with a grid in your finished work.

ORTHOGRAPHIC

An orthographic allows all the faces of a design to be seen. You can also show hidden interior detail using dashed lines. Usually, an orthographic drawing would be made using drawing aids. However it can be useful to sketch a freehand orthographic if there are parts of the design that cannot be seen if drawn in isometric.

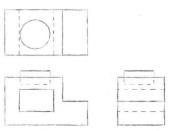

△ A freehand orthographic

SECTIONAL VIEW

You could also show interior detail by sketching a sectional view. This type of drawing can help the designer visualise how parts will go together, as well as show features such as holes through an object. Try and sketch cross-hatching at 45°.

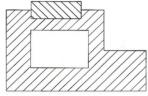

△ A freehand sectional view

SKILLS ACTIVITY

Practising the different types of drawing is the key to mastering them. Study the picture of the clock. Using the most appropriate freehand drawing method, produce notes and sketches for:

- instructions on how to assemble the hands to the clock
- a redesign of the clock in a different style
- a drawing showing three sides of the clock.

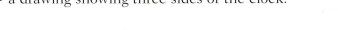

KNOWLEDGE CHECK

Which type of drawing would be used to produce the following:
- a drawing of a torch showing interior detail?
- design ideas of a shop sign in three dimensions?
- a sketch of chocolate packaging showing how a plastic tray is inserted?
- a sketch of a children's playground?
- a sketch of a living room showing the placement of furniture in the room? (5)

2.4 Drawing basic shapes

LEARNING OBJECTIVES

By the end of this unit you should:
- ✓ be able to construct a variety of regular and irregular polygons accurately
- ✓ be able to construct circles accurately
- ✓ be able to construct ellipses accurately
- ✓ be able to construct tangents and tangential arcs accurately.

DRAWING BASIC SHAPES

Being able to accurately construct a variety of basic shapes, such as circles, ellipses and polygons, is important for producing accurate drawings of designs. These days, computer-aided design is used to create shapes to a high degree of accuracy as computers can work out angles and dimensions, allowing quick and efficient production of shapes.

REGULAR POLYGONS

A polygon is a shape with three or more sides. A **regular polygon** has sides of equal length and internal angles of equal size. To find the central angle you can divide 360° by the number of sides.

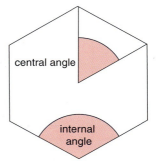

△ A hexagon showing the internal and central angles.

SKILLS ACTIVITY

Copy and complete the table showing some regular polygons and their central angles.

Number of sides	Name	Central angle (°)
3	triangle	120
4	quadrilateral (square)	90
5	pentagon	?
6	?	60
7	heptagon	?
?	?	45
9	?	?

Drawing a hexagon using a compass and a protractor

1. Draw a circle with the same radius of your intended hexagon.

2. Place the protractor on the circle lining up the centre. Mark out 60° on the circle and draw a line from the centre.

3. Rotate the protractor so it is lined up with the line you have drawn. Repeat the previous step.

4. Repeat all the way round the circle.

5. Join up the areas where the lines intersect the circle.

TOP TIP

Another way of drawing a polygon is by using a compass to measure the distance between the first two lines, and marking out the remainder of the points of the polygon on the circle before joining them up.

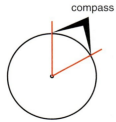 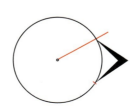 

△ You can use a compass instead of a protractor to mark out the points after the first two lines have been drawn.

IRREGULAR POLYGONS

An **irregular polygon** has sides of different lengths and internal angles of different sizes. A four-sided irregular polygon is called a quadrilateral. A square, hexagon, kite, diamond and parallelogram are all examples of quadrilaterals. Below are more examples of some irregular polygons.

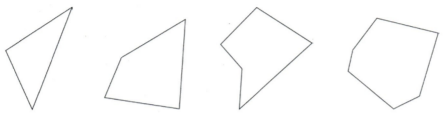

△ Irregular polygons have unequal internal angles and side lengths.

CIRCLES AND ELLIPSES

You can draw a circle by drawing lines of equal distances and angles and connecting the end points. Draw six lines as a minimum and increase the number of lines for greater accuracy.

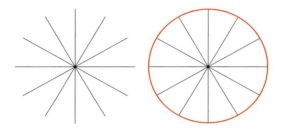

To draw an **ellipse**, start with the same angled lines as you would for a circle and draw two circles the diameters of the intended minor and major axis of the ellipse. Where the angled lines intersect the smaller circle, draw horizontal lines. Draw vertical lines where they intersect the larger circle. Make a point where these two lines intersect, then join up all of the points.

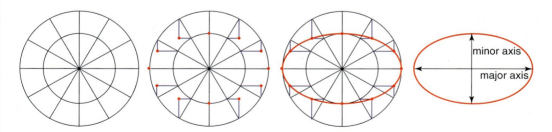

Another way you can draw circles and ellipses is to use a template or stencil. However, these limit the sizes you are able to draw.

Drawing an ellipse using a trammel

A third way of drawing an ellipse is by using a trammel. A trammel can be made from a folded piece of paper or strip of card. This is an easy and accurate method of drawing an ellipse.

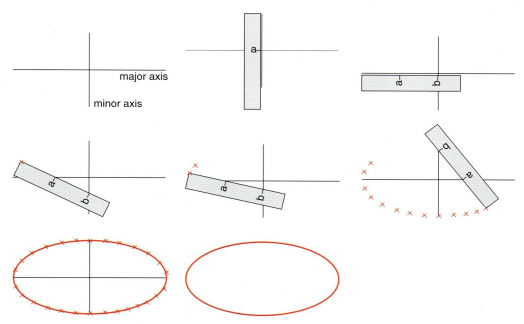

△ Ellipses can also be drawn using the trammel method.

1. Draw two intersecting lines the lengths of the intended diameter of your minor and major axes.
2. Lay a folded piece of paper or strip of card against the minor axis with the corner of the card touching the top of the minor axis.
3. Mark (a) on the card where the axes intersect.
4. Rotate the card and mark (b) where the axis intersect.
5. Keeping (a) somewhere on the major axes and (b) on the minor axes make a mark at the top corner of the card.
6. Repeat this several times around the axes.
7. Join up the marks for an ellipse.

TANGENTS AND TANGENTIAL ARCS

A **tangent** is a line that touches a circle or curve at a certain point.

 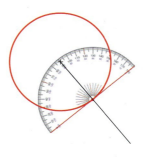  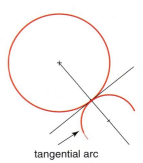

tangent   tangential arc

△ A tangent and tangent arc can be drawn using a protractor.

To draw a tangent:

1. Draw a circle with a point where you want the tangent.

2. Draw a line from the centre through the point.

3. Place the centre line of your protractor on this line and the bottom of the protractor just touching the circle.

4. Draw the tangent.

5. You can also use these lines to draw a tangential arc. Draw a perpendicular line from the centre of the circle through the part of the circle you want your tangential arc to be. Set your compass to the correct radius and draw the tangential arc, making sure the point of your compass is on the perpendicular line you have drawn.

KNOWLEDGE CHECK

1. Explain two different ways of drawing a circle. (4)
2. Explain two different ways of drawing an ellipse. (4)
3. Explain the difference between a regular and irregular polygon. (3)

KEY TERMS

ellipse: an oval shape like a flattened circle. The larger diameter is the major axis and smaller diameter the minor axis. It is also a shape produced when a cone is cut through at an angle.

irregular polygon: a shape with three or more straight edges of different lengths and with different internal angles

regular polygon: a shape with three or more straight edges of the same length and with equal internal angles

tangent: a line that touches a circle or curve at one point but does not intersect it

2·5 Developments

LEARNING OBJECTIVES

By the end of this unit you should:
✓ understand what a development is and how to construct it
✓ be aware of a range of different developments
✓ be able to add constructional details such as tabs and flaps.

A **development**, or **net**, is a three-dimensional shape that has been folded out and flattened. Designers should be able to visualise a development from a shape and draw an accurate development to scale from an orthographic or pictorial view. Being able to construct a development is an important first step in producing packaging. To help produce developments, computer-aided design can be used to produce secure nets for packaging which are interesting to look at, and easy to assemble and open. Machines such as die cutters can be used to produce developments in quantity before they are assembled and used as packaging.

TOP TIP

The first step in producing a development from a pictorial view is to visualise unfolding the shape. In your head, begin with the side that is flat on the surface and unfold each side until it is flat. It helps to make a quick sketch of the development before you draw it accurately or to scale. Remember to consider how many sides the shape has and position the first face carefully on the paper so you fit it all in.

SKILLS ACTIVITY

Here are the pictorial views of three shapes. Sketch the development of each shape. Estimate all sizes.

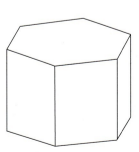
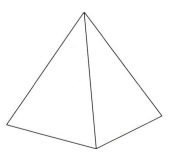

TRUNCATIONS

A **truncation** is a shape that has been cut at an angle other than 90°. A truncation can be drawn from an orthographic projection.

PACKAGING DEVELOPMENTS AND CONSTRUCTIONAL DETAILS

Designers need to be able to draw a development of packaging or a container, as well as add constructional details such as **flaps** and **tabs** for assembly. Dust flaps are usually added either side of the lid to prevent dust from entering the packaging. A tuck flap appears on the front of the lid and helps keep the lid secure. Glue flaps or tabs are added to any sides that need to be secured to another side to hold the packaging together. Edges which need to be folded are show as a dashed line. More detail on flaps and tabs can be found in the materials and modelling section in unit 2·12.

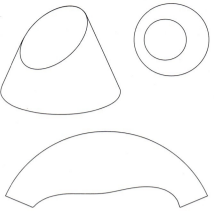

△ Truncations are three-dimensional shapes that have been cut.

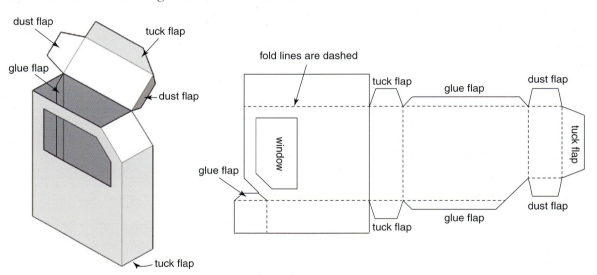

△ Packaging nets have flaps and tabs added to allow for easy assembly and security.

> **DESIGN IN ACTION**
>
> In design and manufacturing, the production of packaging such as milk cartons is a fully-automated process. After the card is printed, it is often sealed with a thin layer of plastic to protect the printed surface and make it resistant to water and scratches. The card is then fed through a series of rollers which crease and cut the card into the correct shape. Glue is then added to any relevant tabs by the machine, which finally folds each packaging and assembles it into the correct shape, before adding the milk. This automated production process needs to be fast, efficient and accurate to ensure a large quantity is made to a high standard in a short space of time.

SKILLS ACTIVITY

Here are two pieces of packaging.

Sketch the developments of both pieces of packaging. Add any relevant flaps and tabs to ensure the packaging is secure and can be opened easily. Estimate all sizes.

KNOWLEDGE CHECK

What is the purpose of adding the following to a development?
1. dashed lines (1)
2. glue tabs (1)
3. dust flaps (1)
4. tuck flaps (1)

> **KEY TERMS**
>
> **development/net:** a folded out net of a container, packaging or a three-dimensional shape, for example a piece of packaging that has been flattened and folded out
>
> **flaps:** parts of a development that stick out, used for tucking sections (such as lids) inside securely
>
> **tabs:** parts of a development that stick out, and which allow sections of it to be glued together
>
> **truncation:** a cylinder or prism that has a part of it sliced away

2·6 Enlarging and reducing

LEARNING OBJECTIVES
By the end of this unit you should:
- ✓ be able to enlarge and reduce circles
- ✓ be able to enlarge and reduce irregular and regular polygons
- ✓ be able to enlarge text
- ✓ be able to enlarge and reduce using one-point perspective.

When designing it is important to be able to accurately enlarge and reduce shapes to a given scale. You can use the following techniques to enlarge a variety of shapes. To reduce a shape you can simply reverse the steps.

ENLARGING CIRCLES AND REGULAR POLYGONS USING A COMPASS

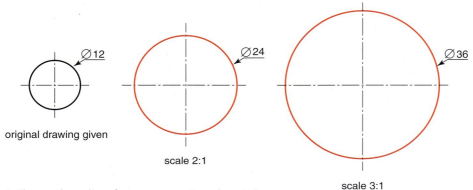

△ Change the radius of your compass to scale a circle.

In the original drawing the diameter of the circle is 12 mm. Therefore, to enlarge to twice the size (scale 2:1), set your compass to radius 12 mm to draw a circle with a diameter of 24 mm. Set it to 18 mm to draw a 36 mm diameter circle, or scale 3:1.

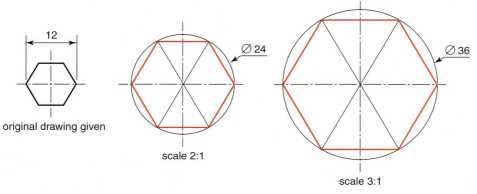

△ Change the radius of your compass to scale a regular polygon.

In the original drawing the length of the hexagon is 12 mm. Therefore, to enlarge to twice the size (scale 2:1) set your compass to radius 12 mm to draw a 24 mm circle. Use a protractor to draw lines at the correct angles and join up the points. Set your compass to 18 mm to draw a hexagon at scale 3:1. You can use a compass to step off round the circumference of the circle, as explained in unit 2·4.

ENLARGING IRREGULAR POLYGONS AND TEXT USING A GRID

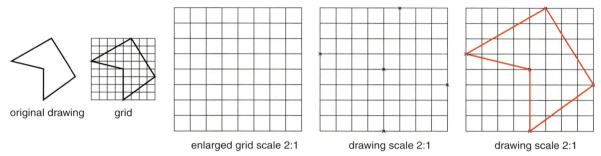

original drawing   grid   enlarged grid scale 2:1   drawing scale 2:1   drawing scale 2:1

△ Use a scaled grid to enlarge text.

1. Draw a box around the original drawing and split it into quarters.
2. Draw a grid accurately around the shape.
3. Enlarge the grid by the scale which you require.
4. Mark on the points of the irregular polygon and then join them up.

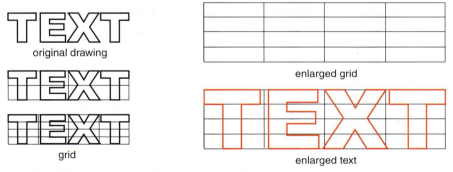

original drawing

grid

enlarged grid

enlarged text

△ Scaled grids can also be used to reduce and enlarge text.

1. Draw a box around the original text.
2. Split the box into quarters the split again evenly to create a grid. Do not worry if the edges of the text do not line up with the grid.
3. Measure the grid, then redraw it to the scale you want. Draw the text inside the grid.

> **TOP TIP**
>
> Make sure the spacing on your grid is an easy number to calculate, such as 5 mm or 10 mm.

ENLARGING AND REDUCING USING ONE-POINT PERSPECTIVE

One-point perspective is a type of three-dimensional drawing that uses a horizon line and a single vanishing point to create a sense of perspective in the drawing.

One-point perspective can be used to enlarge and reduce a shape proportionately.

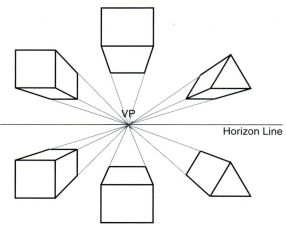

△ A one-point perspective uses a single vanishing point.

1. Draw a line from the corner of the square through the middle of the opposite side.

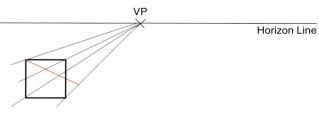

2. Where this line intersects with the bottom vanishing point line is the bottom corner of a new square. Draw this square using the vanishing points as a guide.

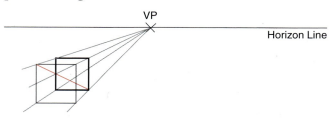

3. Continue drawing lines from the corner of the new square to create boxes at equal distances.

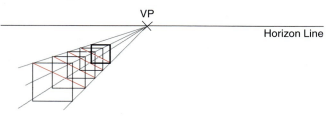

4. You can split the squares into a grid to help reduce more complex shapes such as this H.

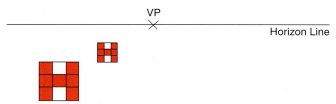

SKILLS ACTIVITY

Imagine that each of these shapes is 50 mm × 50 mm.

1. Enlarge each shape using the scale 2:1.

2. Enlarge each shape proportionately using one-point perspective.

KNOWLEDGE CHECK

1. A sign maker would like you to enlarge a sign displaying the word 'FORTUNE'. Outline the steps needed to enlarge the sign accurately. (5)

2.7 Instruments and drafting aids

LEARNING OBJECTIVES
By the end of this unit you should:
✓ be able to recognise and use a range of different instruments and drawing aids to create accurate drawings with a good standard of graphical representation.

Accuracy and good presentation are essential for communicating ideas to other designers, the client or the manufacturer. Having the ability to select the correct drawing aid is important.

Rulers, protractors and set squares
Use a ruler to measure lines. Protractors are for drawing angles accurately. Set squares can be 30°/60° or 45°/45° and are used to draw these angles accurately.

△ Measuring and drawing tools

Stencils
Stencils can be used to draw circles, ellipses and lettering.

△ A circle stencil

Technical pens
Technical pens have a small nib. They come in different nib sizes in mm, and the correct pen should be selected to correspond with the thickness of the line you are drawing.

△ Drawing pens come in different thicknesses.

Compasses
A compass is essential for drawing circles; simply set the radius of the circle you want to draw. Compass cutters are also available for cutting circles from paper or vinyl.

△ A set of compasses is essential in technical drawing.

Flexicurves
Flexicurves can be moulded to any given curve. They can be used as a curved ruler, to measure and draw curved lines.

△ A flexicurve

Drawing board
A drawing board has a parallel motion rule which moves up and down on the board. This keeps lines horizontal. You can place set squares on this rule to draw lines in 30°/60° and 45°.

△ Drawing boards and clips help with accuracy.

SKILLS ACTIVITY

Make a table with two columns. In the first column list all the shapes you have learnt. In the second column, list all the instruments or drawing aids required to draw these shapes.

SKILLS ACTIVITY

Here is an orthographic view of a keyring.

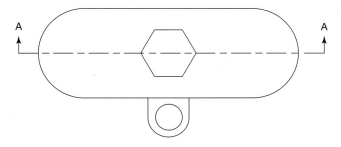

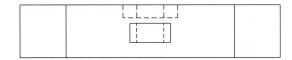 

1. Draw an isometric view of the keyring, using appropriate tools.

2. Draw a sectional view along A–A, using appropriate tools.

KNOWLEDGE CHECK

Which drawing aids and instruments would you select to draw the following?
1. a planometric drawing of the interior of a room (1)
2. an isometric drawing of a mobile phone (1)
3. an orthographic drawing of packaging in the shape of an octagonal prism (1)
4. a sectional drawing of a torch (1)
5. a freehand two-point perspective drawing (1)

2·8 Layout and planning

LEARNING OBJECTIVES

By the end of this unit you should:
- ✓ be able to set up an A3 sheet for drawing in orthographic, isometric and perspective
- ✓ be able to plan a drawing.

LAYING OUT AN A3 SHEET

Before you start drawing it is important to lay out your A3 sheet correctly. You should draw three lines at the bottom of the page to insert information about the drawing, such as the title of the drawing, the scale and the date. This helps to keep your work neat and organised.

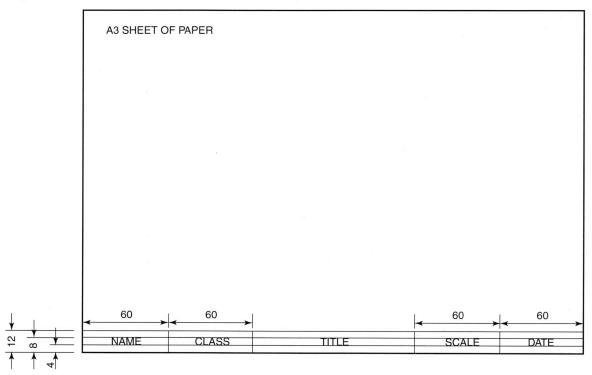

1. Attach your A3 sheet of paper to the drawing board using clips.
2. Draw three lines across the bottom using the parallel motion tool. The lines should be 4 mm apart.
3. Draw two lines from the left and two from the right, 60 mm apart.
4. Write in the details as shown.

Laying out for an orthographic drawing

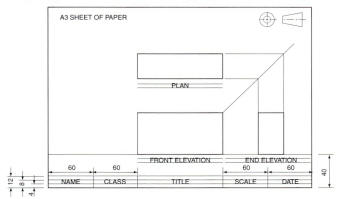

1. Draw the third-angle or first-angle projection symbol in the top right of the page.
2. Draw a base line 40 mm up from the bottom of the page.
3. Draw a box in the middle which is the total height and length of the object you are drawing.
4. Project a 45° line from the corner of this box.
5. Change the views around if you are drawing first-angle projection.

TOP TIP
Construction lines should be light so they do not get confused with the main part of the drawing. In design it is good practice to leave the construction lines to show how the drawing was created.

Laying out for an isometric drawing
1. Draw a centre line.
2. Mark on the centre line near the bottom and draw a crate which is the total length, height and width of the object you are drawing.
3. Draw your object inside this crate.
4. Add dimensions if necessary.

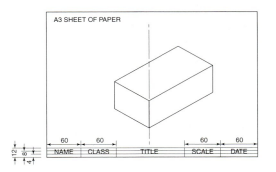

Laying out for an estimated two-point perspective drawing
1. Draw a horizontal line near the top of the page.
2. Mark two vanishing points towards the ends of the line.

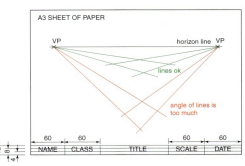

TOP TIP

When drawing construction lines from the vanishing points, make sure you do not draw them at an angle too far from the horizon line. If the angle of the lines is too large you will not be able to fit the drawing on the page.

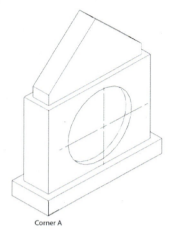 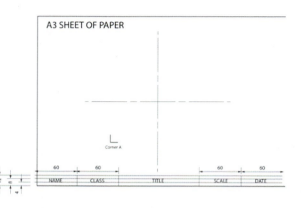

In the design industry, designers are sometimes given lines to work from such as centre lines or corners. This acts as a guide to help the designer fit the drawing on the page.

SKILLS ACTIVITY

Here is a pictorial view of a leaflet holder.

Set up A3 pages and draw the following:

1. an orthographic first-angle projection using the scale 1:2
2. an isometric drawing using the scale 1:2
3. an estimated two-point perspective drawing (estimate all sizes).

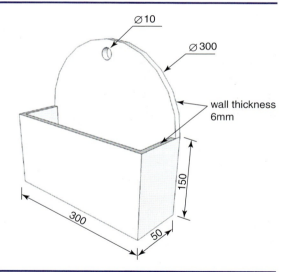

KNOWLEDGE CHECK

When setting up an A3 page for drawing:
1. Where on the page should the projection symbol be? (1)
2. What distance apart should the lines at the bottom of the page be? (1)
3. What two things do you draw before you begin drawing an isometric drawing? (2)
4. How can you ensure you fit a two-point perspective drawing on your page? (1)

2·9 Presentation

LEARNING OBJECTIVES
By the end of this unit you should:
- ✓ be able to use the thick- and thin-line technique to enhance a drawing
- ✓ understand light source and its effect on light and shade
- ✓ use rendering to emphasise three dimensions, textures and materials.

PRESENTATION IN DESIGN
Having the ability to present design proposals in a clear and attractive way can help designers communicate their ideas effectively to clients. Certain techniques such as thick- and thin-line and **rendering** can enhance the quality of drawings. In the design studio, CAD is often used to show **light sources** and materials used. The computer program can then render the drawing to give an extremely life-like appearance to the object.

THICK- AND THIN-LINE TECHNIQUE
By using different line widths it is possible to give a three-dimensional drawing emphasis. You can use technical fine line pens with different widths or different pencils to achieve this effect.

TOP TIP
Look at this isometric cube and imagine a spider on the top face. Now imagine the spider crawling over one of the edges. Can you still see the spider? If so then the line that the spider has crossed over is thin. If the spider disappears behind, inside or under the cube then that line is thick.

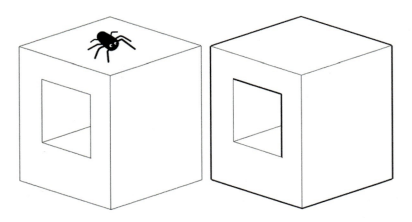

SKILLS ACTIVITY

Sketch the shapes below in freehand. Use the thick- and thin-line technique to enhance your drawings.

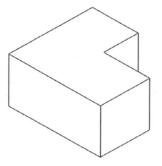 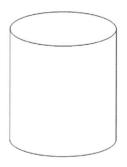 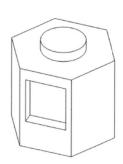 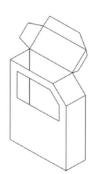

LIGHT SOURCE AND RENDERING

Before rendering a drawing you will need to consider where the light source is coming from. The direction of the light source will affect what tone you need to apply to the object's faces. For cylindrical objects you should use a smooth gradient from dark to light to dark.

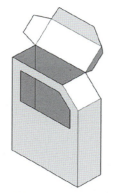 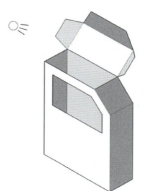 

△ The direction of the light source affects the tones on an object.

TOP TIP

For a smoother finish, use a softer pencil such as a 2B. Turn the paper as you render so the pencil strokes are applied in different directions.

SKILLS ACTIVITY

Practise rendering by shading in the shapes you drew in the previous Skills Activity. Use a light source coming from the top left of the object.

RENDERING MATERIALS AND TEXTURES

Using coloured pencils and an eraser you can achieve different textures to represent materials.

Plastic

Shade the faces taking into account the light source. Use an eraser to make streaks on the faces in the same direction to give a reflective look. Finally, lightly add in the back edges to give a transparent look.

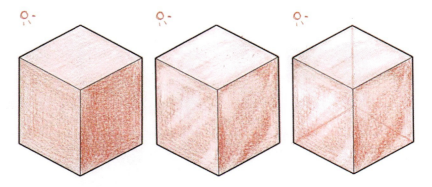

△ Plastic rendering showing a translucent cube with highlights.

Metal

Use a grey coloured pencil to shade in the faces, taking the light source into account. Add some light blue tints. Use an eraser to remove highlights to give a shiny look.

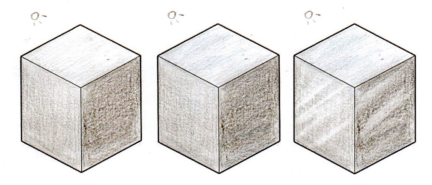

△ Use a grey and light blue pencil to give a metallic effect.

Wood

Use a brown, ochre and yellow pencil to roughly and unevenly shade the faces. Since wood does not have a smooth finish, aim to produce a rough look. Finally, add grain with the brown pencil, loosely following the correct angles on the face.

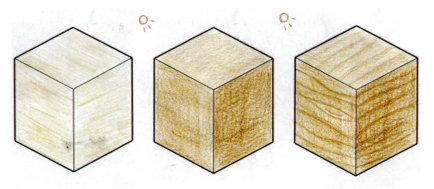

◁ Wood grain can be shown using uneven lines.

SKILLS ACTIVITY

Here is a pictorial view of a toy train. Draw the toy train and render it to give a realistic effect of the materials stated below. Use a light source from the top left to show different tones and finish using the thick- and thin-line technique.

- Render the cylindrical front part in plastic (any colour).
- Render the wheels in metal.
- Render the body in wood.

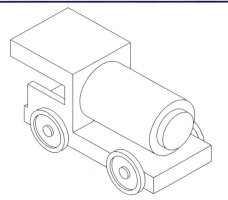

LETTERING AND FONTS

Selecting and producing lettering is a key aspect of graphic design. This is called **typography**. Fonts can be used in headings and titles for their visual impact, or chosen for readability in the main body of a text.

Lettering styles

There are two main types of lettering; serif and sans serif. Serif fonts have tails at the end of the letters and sans serif fonts do not. Sans serif fonts tend to be cleaner and look modern, while serif fonts look more sophisticated and can aid readability in large chunks of text.

ABCDE ABCDE

Tracking and kerning

Tracking is adjusting the space between the letters of a word by equal amounts. This is often done for style reasons. Tracking is often increased in the logos of sophisticated fashion brands as it gives the illusion of wealth and sophistication. Alternatively, some contemporary designers tighten the tracking of the letters so much that they overlap.

Sometimes when words are typed they do not look quite right. In this case **kerning** can be used to adjust the space between pairs of letters to improve the look and balance of the overall word.

AVAST   AVAST

The first typed word shows inconsistency in the letter spacing between the As and the V. The second is the same word with kerning applied between the As and the V.

SKILLS ACTIVITY

Find out how to adjust the tracking and kerning in a piece of computer software that you use. For example, in Adobe Photoshop and Illustrator select the letters, hold alt and press the right and left cursor keys. Experiment with adjusting the letter spacing in some text.

Applying text

There are numerous ways to apply text to a graphic product. Computer generated text can be printed and applied with spray glue. You can also use a CAM plotter to cut text from vinyl and use transfer vinyl to lift it and apply it to your product. Two other methods include dry transfer (Letraset®) and stencils.

KNOWLEDGE CHECK

1. Explain why presentation is important in design. (3)
2. Outline three ways you can improve the presentation of your design ideas. (3)
3. When rendering, which type of shape would have a gradient applied? (1)
4. What type of pencil should you use for rendering? (1)

KEY TERMS

kerning: the process of adjusting the space between pairs of letters in a word

light source: the point or place from which the light is coming

rendering: the process of shading a drawing using light and dark tones

tracking: the process of adjusting the space between all the letters in a word

typography: the art of designing, placing and adjusting lettering to make it readable, legible and attractive for printing

2·10 Data graphics

LEARNING OBJECTIVES
By the end of this unit you should:
- ✓ be able to produce line, pie and bar charts from data provided
- ✓ produce sequence drawings and flowcharts.

GRAPHS AND CHARTS
Graphs and charts are a good way of visually representing data. You should be able to select the most appropriate graph or chart to represent given data.

Bar charts
Bar charts are used to represent data that is discrete or in categories (e.g. number of sales in a particular year, eye colour, shoe size). The **x-axis** shows the values or category labels and the **y-axis** shows the amount. Bar graphs can be drawn in two dimensions, or three dimensions for visual impact.

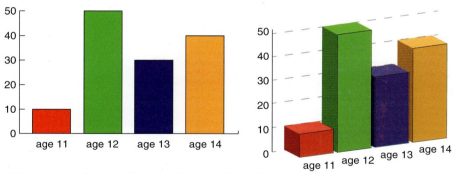

△ Bar charts can be two-dimensional or three-dimensional.

TOP TIP
When drawing bar and line graphs, use a ruler or grid paper to measure out the equal distances between the data range on the x-axis and y-axis.

Line graphs
Line graphs are used to show data that is continuous and often show changes over time. In this case the x-axis shows the time scale and the y-axis shows the continuous variable. Line graphs can also be used to show a trend in discrete data over time (e.g. number of sales in successive years). You may need to use a **key** to label the information.

▷ An example of a line graph

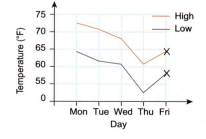

Pie charts

Pie charts can be used to compare percentages or parts of a whole. Again, a key is used to label the sections of the chart.

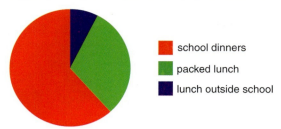

- school dinners
- packed lunch
- lunch outside school

△ Drawing a pie chart is a useful way of comparing data. In this example it is easy to see that the least popular lunch option for students is eating outside school.

TOP TIP
When drawing a pie chart, use a protractor to split the chart into the correct proportions.

SKILLS ACTIVITY
Draw a suitable graph or chart to represent the following:

1. A school of 1150 students is organising a sports festival and have offered three sports options to students. 220 choose basketball, 570 choose baseball and 360 choose swimming.

2. A school wanted to record the change in the number of students who walk, cycle and take the bus to school over a three-year period between 2011 and 2013. Below is a table representing the data.

	2011	2012	2013
Walk	145	258	340
School bus	387	270	210
Cycle	38	42	20

TOP TIP
Graphs and charts can be automatically generated from data using CAD or spreadsheet software.

SEQUENCE DRAWINGS AND FLOWCHARTS
Sequential drawings and flowcharts can be used to show stages in a process. They can also be used to explain how to do something step by step.

Sequence drawings

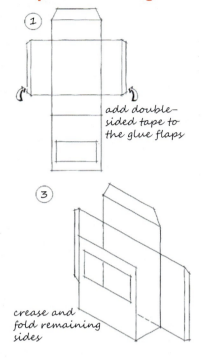
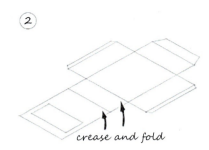
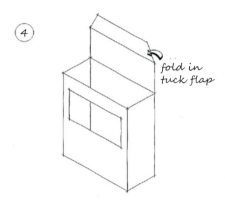

This sequence drawing illustrates how to construct packaging. The drawings are clear and labels and arrows help to inform the viewer. They are clearly numbered.

Flowcharts

A flowchart shows a process in a sequential manner. It is often used in design to give information on how to make something. A flowchart always runs from top to bottom and consists of a series of shapes with connector arrows. The shapes used to show different functions during the process are as follows:

- A rounded shape is called a terminator, and is used at the start and end of the process.

- A rectangle is used to show a stage in the process.

- A diamond shape is used for a decision or quality check. There should be two arrows leading from this box to other parts of the process. This is called a feedback loop.

- A parallelogram is used for input/output. This stage indicates something which is added or removed from the process.

△ Flowchart symbols

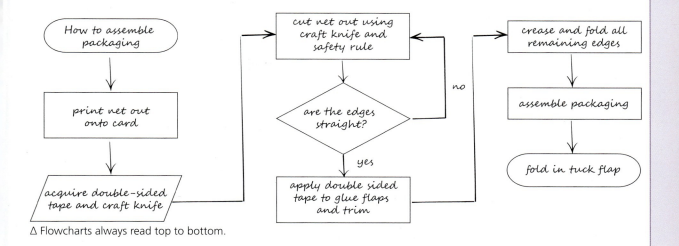

△ Flowcharts always read top to bottom.

SKILLS ACTIVITY

1. Complete a sequence drawing showing how to lay out an A3 page for an orthographic third-angle-projection drawing.
2. Using a flowchart, explain how to draw an ellipse using the trammel method.

KNOWLEDGE CHECK

A manufacturer of stationery products requires some data analysis of who buys their products and which products are most popular. Answer the following questions:

1. Which graph or chart should be used to represent the increasing or declining popularity of pencil sharpeners over the period of five years? Justify your answer. (2)
2. Which graph or chart should be used to represent the proportion of different age groups that purchase protractors? Justify your answer. (2)
3. Which graph or chart should be used to compare the sales of five different stationery products? Justify your answer. (2)
4. Give an example of how an input/output box would be used in a flowchart. (2)
5. What is the purpose of a feedback loop? (1)

KEY TERMS

key: a list of explanations for the different categories used on a chart or graph

x-axis: the horizontal line of a graph

y-axis: the vertical line of a graph

2·11 Reprographics

LEARNING OBJECTIVES
By the end of this unit you should:
✓ understand a range of commercial printing methods.

Commercial printing has come a long way since the earliest forms of woodblock printing in China around 200 CE. Modern commercial printing methods allow hundreds of thousands of copies of an image or text to be printed in high **resolution** in a short space of time using the **process colours** CMYK (Cyan, Magenta, Yellow and Key (black)). Different methods of printing are used depending on the effect and the **substrate** (material) to be printed on. As well as printing, many reprographic systems allow for cutting, creasing and stapling.

There are four primary methods of printing:

OFFSET LITHOGRAPHY
Offset lithography is a widely used form of printing where the substrate is fed through four different rollers, each printing one of the four process colours. Sometimes a fifth roller is used to print a special printing effect such as varnish. Offset lithography is used for medium to long print runs to print items such as leaflets and brochures.

GRAVURE
Gravure is a high quality form of printing where an etched **printing plate** fills up with ink and deposits the colour as small dots. A separate roller presses the material to be printed onto the printing plate. Gravure is used to print high quality items such as company brochures.

FLEXOGRAPHY
Flexography is a form of relief printing, which uses a flexible relief plate with the image to be printed raised up off the non-printing surface. This allows for a variety of substrates to be printed on, such as plastic and metallic films. Products that are printed using flexography include plastic bags, disposable cups and chocolate wrappers.

△ Flexography is used to print onto different materials.

SCREEN PRINTING
This form of printing uses a porous mesh with fine holes which ink can pass through. In screen printing, a 'squeegee' is used to scrape the ink through the exposed holes in the mesh and onto the substrate. Screen printing is used to print onto fabrics, which make it popular for t-shirt printing.

△ A squeegee is used in screen printing.

> **DESIGN IN ACTION**
>
> When an image comes off the printing press it does not look like the finished article. There are a series of symbols surrounding the image. These are called printer's marks and are used to check the quality of the colour and the alignment of the image. If hundreds of thousands of copies are being made of a newspaper or leaflet, one small mistake could be very costly for the printer. This is why the prints need to be checked thoroughly for any problems.

TOP TIP

When starting to construct a design using CAD, it is important that you consider the final medium in which it will be viewed. If your design will only be seen on a computer screen, such as a web page, make sure you set the resolution of your document to 72 dpi (dots per inch). If you are intending to print the graphic, you should set the resolution to 300 dpi for a good print quality.

SKILLS ACTIVITY

Look at some printed items around you. These could be forms of packaging, books, clothing, wrappers or posters. There are examples everywhere. Consider the quality of the print, how many would have been printed, and the material that has been printed on. What type of printing do you think was used for these products? Explain your answers.

KNOWLEDGE CHECK

Which commercial printing method would be used to print the following:
1. a foil bag of snacks?
2. a newspaper of 200,000 copies?
3. a tea towel?
4. a postage stamp? (4)

> **KEY TERMS**
>
> **printing plate:** a sheet which is wrapped round a roller and used to transfer the image to the substrate
>
> **process colours:** the four colours that are used in printing: Cyan, Magenta, Yellow and Key (black)
>
> **resolution:** the quality of images measured in dpi (dots per inch). The higher the dpi the better the quality of the print, but the larger the file size used to store the image.
>
> **substrate:** the material that something is printed on

2·12 Materials and modelling

LEARNING OBJECTIVES
By the end of this unit you should:
- ✓ understand the reasons for modelling
- ✓ understand how to use modelling to an appropriate scale
- ✓ understand why scale drawings enable a visual model to be made
- ✓ understand the properties and uses of a range of modelling materials
- ✓ recognise a range of modern adhesives and their uses.

THE IMPORTANCE OF MODELLING IN DESIGN
Modelling is essential in the design of graphic products. It can take many forms and occur at any stage in the design process. Some designers will visualise ideas early on using quick sketch models and **prototypes**. More accurate **scale models** are often made later on in the design process for testing or presenting ideas to a client.

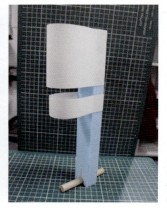

◁ Modelling can be used to visualise and test ideas at all stages of the design process.

TOP TIP
Use modelling at all stages of the design process to help visualise and test your ideas. Do not forget to record your modelling: taking plenty of photos helps you keep track of the progress of your work and shows your thought process to others.

MODELLING TO SCALE
In most cases it is important to model to scale. This means either creating the model so it is roughly the same dimensions as the final design will be, or in proportion to the final design. Sometimes the product being designed is large in scale, such as an exhibition layout. In this case it would take far too much time and resources to model

ideas to the actual size of the exhibition, so smaller scale models would be produced. These scale models would take less time and use less materials, yet still enable the designer to visualise their ideas effectively.

Using scale drawings for modelling

In the design industry, designers often send designs to a manufacturer who will then create a high quality prototype using a variety of commercial production techniques. This saves the designer time as they do not have to produce the model themselves. It is important, therefore, that accurate scale drawings are given to the manufacturer so that an accurate model can be made. These drawings will usually be an orthographic projection and will include information about scale, materials and finishes. If you are making a prototype yourself, an orthographic drawing is still essential in producing a successful prototype.

MODELLING MATERIALS

There are a variety of materials that can be used for modelling graphic products. It is important to consider the properties of materials when selecting which material to use. Consider what it is you are designing. Is the product 'blocky'? Does it have curved or straight corners? Is it transparent or opaque? Does it need to open and close? Considering the purpose of your intended design will allow you to choose the most suitable materials for modelling.

Paper and card

Paper and card are available in different weights and finishes. The weight of the paper or card is measured in gsm (grams per square metre) which affects the thickness of the material. Paper and card are easy to cut, crease and shape. They are also easy to print onto. Paper can be used for sketch modelling, but it is not durable enough to be used in a final model. Card is better for producing a final model as it is stronger. Card and paper are available in a coated (glossy) or matt finish.

Corrugated card and plastic

These materials are impact resistant due to the corrugated layer in the middle, and are often found in packaging for delicate items like electronic equipment or glassware. These materials can be bent and curved one way but not the other and are good for creating designs with curved edges.

Styrofoam and foam board

Styrofoam and foam board are made from expanded polystyrene. This material is found inside packaging to help protect the contents as it is impact resistant. Styrofoam is soft so it can be easily cut using a scalpel and can be shaped using sandpaper. Foam board comes as sheets and consists of a layer of foam between two layers of card.

TOP TIP

Styrofoam reacts badly with some types of spray paint. If you wish to spray paint a styrofoam model, make sure you add a protective coating of filler first.

Self-adhesive vinyl

This material usually comes in rolls or sheets and can be found in a wide range of colours, glossy and matt. It is made from a plastic called vinyl and has an adhesive backing so can be applied onto products. It can be easily cut using scissors or a craft knife. However, it is often cut using a machine such as a CAM plotter. Vinyl can be used to model the placement of two-dimensional graphics onto a graphic product.

Thin plastic sheet (acetate)

This thin plastic sheeting has many of the same properties as paper or card. It can be easily cut, creased and bent. It is usually used as a modelling material due to its transparent properties and is therefore good to use for packaging where the contents need to be seen.

Smart materials

Smart materials change shape or appearance due to changes in the environment such as heat, light or electricity. Polymorph is made of small plastic granules that soften at low temperatures and are usually added to hot water, then shaped like plasticine. As the polymorph cools it hardens. Polymorph is excellent for modelling for ergonomics as it can be moulded to the shape of parts of the body. Shape memory alloys (SMA) can be shaped and then when heat is applied they return to their original shape. Thermochromic materials change colour when heat is applied and are used in products that change colour to warn the user when something is hot.

PERMANENT AND TEMPORARY JOINING METHODS

When designing graphic products, designers need to decide how the product will be assembled. Permanent joining methods require the application of adhesives onto tabs to hold together the assembled edges into the required shape. If these tabs are taken apart the material is likely to rip or be damaged.

△ Packaging is designed to be assembled.

Temporary joining methods use slots, flaps and tabs and can be assembled without the use of adhesives. This means the product can be flat-packed for transportation for the user to assemble before use.

Permanent joining methods

Glue can be used to hold together packaging and displays using glue tabs. Common types of glue used are glue sticks, hot glue and double-sided adhesive tape. Glue sticks are easy to use but are not strong. Hot glue is good for joining thicker materials. Double-sided tape gives good strength but care must be taken to ensure a quality finish.

> **TOP TIP**
>
> Spray adhesive is weak to allow the repositioning of two-dimensional artwork on a flat surface. Because of this weakness it should not be used to assemble packaging. Use double-sided adhesive tape for permanent joining of glue tabs on packaging. To ensure a good quality finish, cover the whole of the tab with the tape making sure that it overlaps the tab. Turn the tab over and trim the excess tape. Peel off the backing of the tape and stick the tab to the packaging.

Temporary joining methods

These joining methods mean that the packaging can be easily assembled without the use of adhesives. A good example of this is a pizza box. These boxes are supplied to pizza restaurants as flat sheets. The restaurant then assembles the packaging when required by folding over the edges and slotting them into the base.

In the pizza packaging below the front edge folds over twice, overlapping the side tab, and slots into the base. This fold-over locking flap keeps the sides upright and the packaging structurally stable. There is no glue required for this packaging.

△ Pizza boxes use fold over locking flaps.

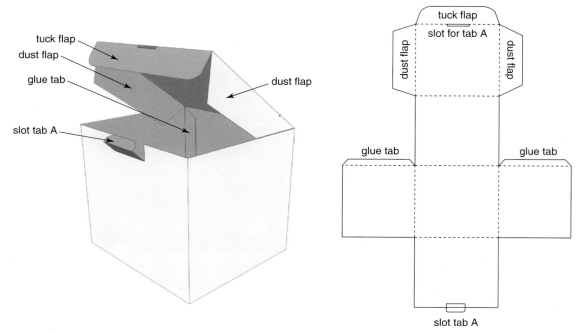

△ Tuck flaps and dust flaps are often added to packaging for security.

In this packaging a tuck flap tucks into the packaging to keep the lid down. There is a slot tab on the front of the box with slots into the lid to hold it in place. The dust flaps add extra reinforcement and prevent dust from entering. Two glue tabs hold the package together.

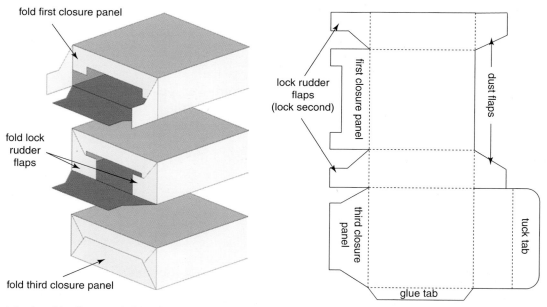

△ Lock rudder flaps can hold a base securely without glue.

In this packaging lock rudder flaps act as a base-locking mechanism, and when assembled in the correct order can securely hold the base of the packaging together. A single glue tab connects the sides.

Arrow tabs can slot in to hold two faces together.

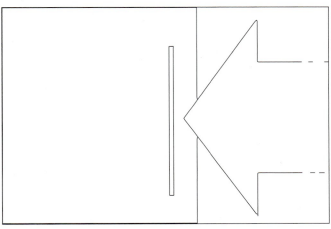

△ An arrow tab and slot

SKILLS ACTIVITY

Find three items of packaging at home. Take them apart so they are flat and consider the following questions:

1. What materials is it made from? What properties make them suitable for this kind of packaging?
2. How is the packaging assembled? Are there any flaps or tabs? How does it lock together as a package?
3. What adhesives have been used to hold the package together?

KNOWLEDGE CHECK

1. Which modelling material is best for the following?
   - modelling a grip for testing the ergonomics of a hairbrush
   - a quick sketch model of a pop-up card
   - modelling a computer mouse
   - a scale model of a display stand (4)
2. Which material and joining methods are best for constructing the following?
   - a pizza box
   - a perfume box (2)

KEY TERMS

prototype: a first trial model of a design

scale model: a model that is either the same size as the intended design or is proportional to the final product

2·13 ICT

LEARNING OBJECTIVES
By the end of this unit you should:
- ✓ understand that digital images can be captured and stored on a computer
- ✓ understand that different types of image manipulation programmes can be used to alter images
- ✓ understand that computers can be used to output to a range of devices
- ✓ be aware of the use of computer-aided design (CAD) and computer-aided manufacture (CAM) in design.

ICT IN DESIGN

The use of computers has become an essential part of design. Computers are used throughout the design process, from capturing, creating and manipulating images and text when developing a design, to the manufacture of a finished product. ICT in design increases productivity, allows for quick changes to be made to designs, and enables effective communication of designs to clients. Designs can be shared through email or collaborated on using cloud computing, allowing many individuals from around the world to take part in the process.

TOP TIP
It is important that you create your own images for use in your designs. Most images found on the internet are not only copyrighted, but are also low resolution. You can use a digital camera or smartphone to capture your own images. These will be better quality and you will have ownership over the images. Alternatively, you can scan your own artwork using a scanner. There are also websites available which allow you to download free fonts for use in your designs.

MANIPULATING IMAGES USING CAD

When using CAD, designs can be altered easily and quickly. There are two main types of image manipulation programs that are used in graphic design; **bitmap graphics** and **vector graphics**.

Bitmap graphics
Used for photographic type images, bitmap images are made up of dots called pixels. The amount of dots per inch in an image is called resolution, the more dots per inch (dpi) the higher the resolution. For screen graphics such as websites the dpi should be set at 72 dpi, for printed images it should be 300 dpi. Bitmap images lose quality if they are scaled up. The most commonly used bitmap software in the design industry is Adobe Photoshop.

△ A bitmap image of a dog at 20 dpi, 72 dpi and 300 dpi. Notice how the quality of the image changes with an increase in dpi.

Vector graphics

Vector graphics programs do not use pixels to create an image. Instead the computer uses mathematical information to work out the size, shape, colour information and outline (stroke) of an object. Commonly-used vector software includes CorelDRAW and Adobe Illustrator.

If you wanted to create a green hexagon with a black stroke, the vector graphics program will understand that a hexagon has six points at six different co-ordinates and that all the internal angles are equal. It will fill the hexagon with a green colour of a certain CMYK value and outline it with a black stroke of a certain point size.

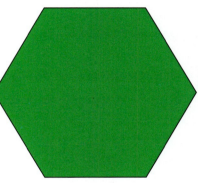

△ Vector graphics use lines, curves and fills.

TOP TIP
The main advantage of a vector graphics program is that images can be scaled without any loss of quality. This makes it great for logos that need to be applied to a variety of different products of different sizes and shapes.

OUTPUTTING DESIGNS USING CAM

As well as printing images, vector designs can be output to **computer-aided manufacturing (CAM)** machines, allowing accurate prototypes and products to be made quickly and easily. Identical products can also be batch produced in quantity using CAM. For more information on CAM, refer to Product Design unit 1·6·1.

SKILLS ACTIVITY
A confectionery manufacturer has asked you to design a new logo for a chocolate called 'Rush'. Using suitable vector graphics software and suitable lettering, design and develop the logo.

KNOWLEDGE CHECK

1. Explain the advantages of ICT in design. (2)
2. Explain what is meant by *vector graphics* and name a piece of vector graphic software. (4)
3. Explain what is meant by *bitmap graphics* and name a piece of bitmap graphic software. (4)
4. Explain the benefits of vector graphics software. (3)

KEY TERMS

bitmap graphic: a graphic that is made up of tiny dots called pixels

computer-aided manufacture (CAM): the use of computers to enable the manufacture of products

vector graphic: a graphic that is created through mathematical information

2·14 Manufacture of graphic products

LEARNING OBJECTIVES
By the end of this unit you should:
✓ be able to use hand tools safely and accurately to manufacture graphic products
✓ understand the use of commercial production techniques for forming plastics
✓ understand the commercial processes used to cut, crease and shape materials.

CUTTING AND SHAPING
In the design studio, designers use a variety of tools for cutting, shaping and manipulating materials for modelling and presenting ideas. These tools exist to aid accuracy and allow for the safe production of prototypes, which are essential for the effective communication of design ideas.

Hand tools for cutting
- **Craft knife**
A craft knife or scalpel can be used to cut paper, card and foam board. Make sure the blade is sharp and make sure you use a safety rule. Use a cutting mat to ensure you do not damage the surface below.

△ A craft knife is used for manually cutting materials.

TOP TIP
Using a craft knife successfully takes practice. Make sure the sharp end of the blade is facing down and at a 45° angle to the paper. When cutting fine detail hold the craft knife like you would hold a pencil. Do not try to cut all the way through thicker materials such as foam board or thick card in one go. It will take several passes to cut through these types of materials. If the edge of the material is not cutting cleanly then fit a new blade. Finally, be safe. Use a safety rule to avoid injury.

- **Safety rule**
A safety rule helps you cut straight lines safely. It contains raised edges, which prevent the blade slipping and cutting your fingers.

- **Paper trimmer**
A paper trimmer allows you to trim card and paper accurately, safely and quickly.

△ Safety rules prevent the craft knife slipping.

- **Rotary cutter**

A rotary cutter is a device for cutting curved shapes. It has a sharp blade on a wheel that rotates.

Creasing and folding

Creasing tools and machines come in a variety of shapes and sizes. They are blunter than craft knives, allowing you to create a crease to fold. Folding a creased edge accurately can be done using a straight ruler.

△ A rotary cutter

Commercial processes for cutting and shaping

When graphic products are being manufactured in quantity, cutting, shaping and assembling by hand is too slow. Most commercial printers will have a **die cutting** machine. This machine uses a shaped steel blade called a die. A die cutter can cut or perforate (cut dashed lines) in the material.

A creasing machine contains a blunt edge on a handle. When the handle is pulled down it creates an indent in the material. The material can then be folded using a folding table.

THERMOFORMING

Thermoforming processes use heat to soften a material, usually plastic, to allow the material to be shaped. This can aid the production of graphic products such as plastic packaging.

VACUUM FORMING

Some packaging contains plastic which helps to contain and protect the contents. This plastic, usually PVC or polypropylene, is moulded using **vacuum forming**. This packaging, which has a card back and a see-through moulded cavity, is called **blister packaging**.

△ Blister packaging

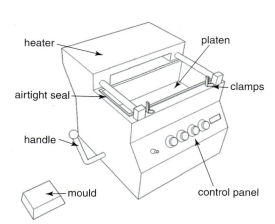

△ Parts of a vacuum former

Vacuum forming process

1. The mould is placed on the platen.

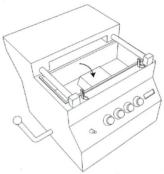

2. A thermoplastic sheet is placed over the airtight seal and clamped above the platen.

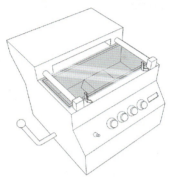

3. The heater is pulled across.

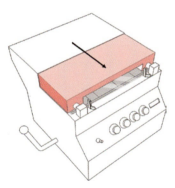

4. The platen is raised so the mould is sitting underneath the plastic.

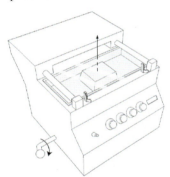

5. The vacuum is turned on and the plastic is sucked over the mould.

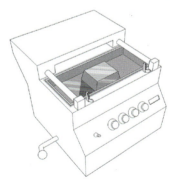

BLOW MOULDING

Blow moulding uses hot air to soften and blow plastic to the sides of a mould. This process is commonly used to produce plastic bottles, shown in the diagram below.

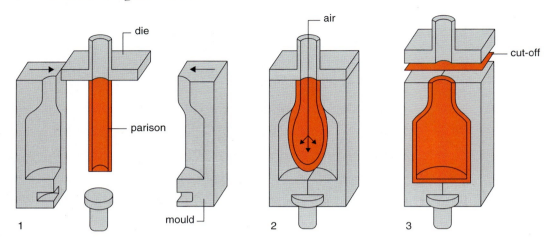

SKILLS ACTIVITY

Practice makes perfect. Acquire a range of materials such as paper, different thicknesses of card and foam board. Use a craft knife to cut through the material without damaging the edges. Practise using a safety rule for straight edges and also try cutting curves and circles. Remember, you may need to cut several times before the blade cuts all the way through the material.

KNOWLEDGE CHECK

Here is a net and assembled packaging. The packaging is made from card with a clear plastic window.

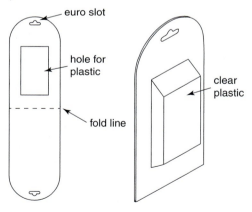

1 Explain the process of:
   a) cutting out the net (4)
   b) scoring and creasing the fold line (2)
   c) forming the plastic window (5)
   d) applying an adhesive and assembling the packaging. (2)
2 Name all tools and equipment required for this process. (2)

KEY TERMS

blister packaging: a type of packaging with a moulded plastic cavity or pocket

die cutting/die stamping: the commercial process of cutting material using a shaped metal die

vacuum forming: the process of heating up a plastic sheet and using a vacuum to suck it over a mould

This section will help you to develop a broad understanding of the materials, techniques, tools, equipment, methods and processes used with resistant materials. Specifically, it aims to support you as you develop your understanding of both the physical and working properties of a range of woods, metals and plastics.

This section is designed to help you to relate your understanding directly to the processes associated with designing and making, so that you are able to apply relevant skills, knowledge and understanding effectively. This is so that following analysis, based on evidence, you can make sound judgements and decisions in relation to material and process choices, in order to plan and carry out work safely and effectively.

You will develop your awareness of the different types of material, including smart and modern materials, and their associated physical working properties. You will learn key terms and understand the processes involved to accurately set and mark out material. You will also learn the correct techniques and be aware of the manufacturing processes involved in shaping, joining and finishing different materials.

STARTING POINTS

- What materials are available?
- What are composite materials?
- How do you decide which type of material is best to use?
- Which tools and equipment are available to cut, shape and join different types of material?
- Which tools and equipment do you use to mark and set out material?
- How do you decide which tools and equipment to use to cut and shape material?
- How do you decide which is the best type of finish to use?
- What is PPE equipment and why is it important to know how to use it?

SECTION CONTENTS

3·1   Types of materials
3·2   Smart and modern materials
3·3   Plastics
3·4   Wood
3·5   Composites
3·6   Metals
3·7   Preparation of materials
3·8   Setting, measuring, marking out and testing
3·9   Shaping
3·10 Joining and assembly
3·11 Finishes

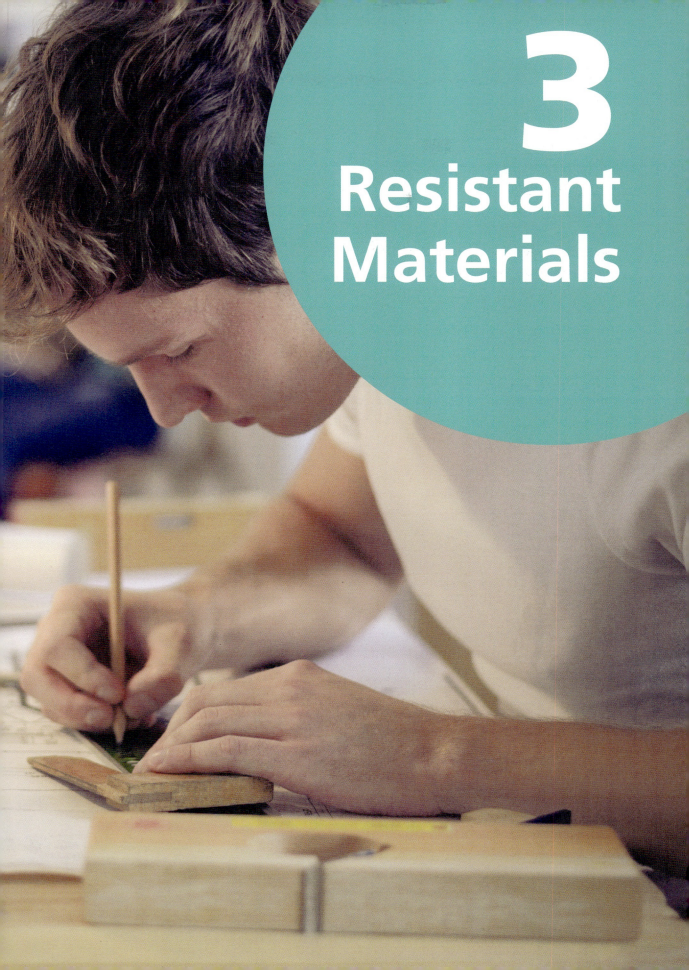

3
Resistant Materials

3.1 Types of materials

LEARNING OBJECTIVES
By the end of this section you should:
✓ understand the common physical and working properties of materials
✓ understand the common factors that should be considered when selecting materials that are fit for purpose.

WHY DO I NEED TO KNOW ABOUT THE PROPERTIES OF MATERIALS?

Designers, technologists and engineers need to know the properties of materials to inform their selection of suitable materials in order to meet the requirements of a product design specification. Two key concepts are critical in relation to your knowledge of materials:

- how materials are classified, so that you have knowledge of a broad range of materials to choose from when designing, and are not restricted to the materials that you have used
- what the properties of materials are, how they add functionality or make a product aesthetically pleasing, and how they can be worked.

CATEGORIES OF MATERIALS

A basic understanding of categories or types of materials can help you classify and select the materials most appropriate for your design needs. This can help you to take a methodical approach to materials and help when interacting with materials databases. The most common materials are woods, metals and plastics, and composites. The terms smart or modern materials are often used to describe manufactured materials with particular properties, which often react to environmental conditions. These are described in detail in unit 3.2. It is also worth noting that textiles are a classification of materials and, while these might be labelled as 'compliant' rather than 'resistant' materials, some materials can be produced in both a textiles (or fabric) and a solid form, such as the polymer nylon which may be used for garments and engineered mechanical components.

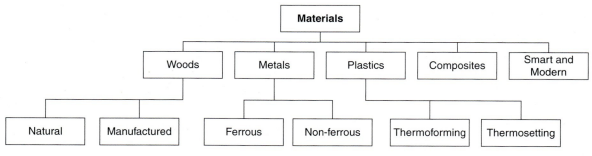

△ A classification table of materials

PHYSICAL PROPERTIES

Physical properties are a major factor when selecting materials for specific users, purposes and conditions. Each physical property describes how a material behaves under specific conditions. Common physical properties are:

- **mechanical:** how a material will react to physical forces, such as how some materials become brittle when repeatedly bent or flexed
- **electrical:** a material's conductive properties including insulation and resistance, such as how a screwdriver's polymer handle (insulator) protects the user from electic shocks through the tip (conductor)
- **chemical:** how a material reacts to other chemicals, such as how metals oxidise when exposed (for example, ferrous metals rusting)
- **thermal:** how a material responds at different temperatures, such as freezing/melting and boiling points, such as the temperature ranges where thermoforming plastics can be shaped
- **magnetic:** most ferrous (containing iron) materials are attracted to magnets or can be magnetised, such as the use of magnetic catches for keeping doors closed
- **permeability:** the degree to which liquids or gases can pass through a material, such as woods that have been varnished having low permeability to liquids and water resistance
- **aesthetic:** the visual sensory properties of the surface/form of a material, such as colour and shine
- **tactile:** the touch or sensory properties of a material, such as texture.

WORKING PROPERTIES

Each **working property** has a technical definition describing how a material behaves when being worked or shaped (see unit 3·9). The most common working properties are:

- **strength:** the ability to withstand a constant external **force** without failing
- **hardness:** the ability to resist wear-and-tear, scratching and dents; usually related to surfaces
- **toughness:** the ability to withstand blows or momentary shocks (live forces)
- **durability:** the ability to resist wear, such as weathering, over time
- **elasticity:** the ability to bend and then to return to its original shape/size
- **plasticity:** the ability to permanently change shape or form, which applies to materials other than polymers
- **malleability:** the ability to permanently change shape, in all directions, while retaining integrity
- **ductility:** the ability to change shape by stretching along its length without snapping.

SELECTING MATERIALS

When selecting appropriate materials for a product it is important to get the design specification right. It is essential to identify the requirements for the user, function or conditions that a product will be exposed to. Once a specification has been drafted, the required physical and working properties can be identified and used to search a materials database or table. For example, a specification for a garden chair might state that the product must be durable for use outdoors in all seasons, with minimum maintenance.

TOP TIP

A common mistake is to rely only on materials that you know and have used before. Understanding how materials are classified, what their properties are, and how to select them will lead to more functional and appealing products. When making a **prototype** (working model/product to test a concept) you might use the materials that are most readily available. However, when writing a manufacturing specification for production, alternative materials should be selected to optimise the efficiency of a commercially viable product, in relation to the cost of manufacture, marketability or functionality.

SKILLS ACTIVITY

Choose a product that you use often and sketch or photograph it. List the materials that you can see. If you are unsure what materials have been used, look in units 3·3 to 3·7 of this book, or go online and investigate the product. Manufacturers' websites often list the materials used. Now select one of the materials that you have identified and list the properties for which it was selected. Give reasons why you think this was the case, including functional, aesthetic, financial and manufacturing factors.

DESIGN IN ACTION

Modern mobile phones and devices are designed around touch screens. Currently most of these devices use special screens that sense electronic charge. Therefore, the glass used needs to have specific properties that enable input from your finger to be picked up by the sensors built into the screen, alongside the visual display. In addition to this the material used needs to be highly transparent, avoid glare and be tough enough to resist normal day-to-day use. The glass used in these devices is designed by material scientists to have particular properties and differs from that used, for example, in windows or spectacles. This is a good example of how important it is that designers, technologists and engineers have a sound knowledge of the properties of materials, so that they can select the most appropriate materials to meet the functional, aesthetic, financial and manufacturing requirements of a product.

KNOWLEDGE CHECK

Design a folding worksurface that could be used by a teenager when doing homework.
1. Communicate your idea using an annotated sketch. (4)
2. Identify all materials used in your design and list the specific physical and working properties which make them suitable. (6)

REFLECTIVE LOG

Think of a product that you have made recently. Write a brief evaluation of the product, focusing on the properties of materials, using these prompt questions:

- What materials did you use?

- Why did you use these materials?

- Would alternative materials be more appropriate if you were to batch or mass manufacture the product? Why do you think this? Make reference to properties of materials.

KEY TERMS

force: a power that produces strain on a material and that changes its shape, or has the potential to change its shape. There are five forces that act on materials: compression, tension, torsion, sheer and bending.

physical property: an observable or measurable characteristic of a material, for instance strength or density

prototype: a working model, in the early stages of development, which, through testing against the specification, is used to help refine the product before beginning final or larger scale production

stiffness: resistance to forces of bending or torsion

stress or strain: a force created by pressure on an object or material, that has the potential to damage it by changing its shape

working property: the way a material behaves when being worked or shaped or while being used within a product. Working properties determine the tools or processes that will be used when making or manufacturing and how a product will function.

3.2 Smart and modern materials

LEARNING OBJECTIVES
By the end of this section you should:
✓ be aware of a range of 'smart and modern materials', including thermochromic materials; polymorph; shape memory alloy (SMA); shape memory polymer
✓ understand the key properties of common 'smart and modern materials'.

WHAT ARE SMART AND MODERN MATERIALS?

'Smart and modern materials' is an umbrella term for a wide variety of materials whose properties alter depending on environmental conditions, such as light, temperature or movement. These materials are designed and manufactured through scientific discoveries, where materials are engineered or altered so that they perform a particular function. They react independently, which is why they are known as 'smart', and some appear to have a '**memory**': for example, deforming and returning to their original shape/structure under certain conditions, such as heat from an electrical current (shape memory alloys and the effect of heat on shape memory **polymers**).

Modern materials also include those manufactured by making changes to the structure at molecular level or using **nanoparticles**, such as the super strong graphene or hydrophobic surfaces that repel water. Some smart and modern materials are inspired by nature; this is called **biomimicry**.

DESIGN IN ACTION Smart and modern materials have numerous applications in health and medicine. A 'stent' is a small mesh tube, made from a shape memory alloy, which is inserted into damaged blood vessels to strengthen or keep them from collapsing. The stent is inserted cold and expands into the desired size with the patient's body heat.

EXAMPLES OF SMART AND MODERN MATERIALS

There are many different materials that can be classed as being smart or modern. Some common examples are shown in the table below.

Name	Properties	Uses
Shape Memory Alloy (SMA) or nitinol	An alloy composed of nickel and titanium, which will (a) when deformed (bent), return to its original shape when heated beyond 90°C and (b) contract when an electric current is passed through it.	SMAs are used in stents, which are tubes inserted into damaged blood vessels. They are inserted cold and expand to the correct size in response to the patient's body temperature, strengthening the blood vessel. SMAs are also used in some spectacle frames or in dentures.
Shape Memory Polymer (SMP)	A variety of polymer (plastics), which deform in response to an external stimulus, such as when an electric voltage is applied or there is a change in temperature.	Potential uses for SMPs are valves or pumps for liquids where mechanical components are not desirable.
Thermochromic inks	Materials with these properties change colour in response to changes in temperature.	Commonly used in forehead thermometer strips and in the food industry to indicate the temperature of a packaged food product.
Photochromic inks	Materials with these properties change colour in response to changes in light level, including ultraviolet (UV) or infrared (IR).	Light reacting sunglasses or spectacles, which darken in response to UV light.
Lenticular sheet	An embossed polymer sheet with optical properties which make a surface appear deeper. There are different kinds of lenticular patterns, which can be used to make images appear to flicker or move.	Can be used to provide visual/aesthetic interest to a surface, such as the illusion of depth, and 3D effects.
Polymorph	Polymorph is available in small pellets which fuse when heated to 62°C and can be moulded, resulting in a tough polymer.	Can be used by designers and engineers to prototype difficult shapes, such as components, joints and handles.
Smart grease	A viscous gel which will uniformly control the movement between two friction surfaces.	Can be used to dampen movement on volume control knobs, slowing down and smoothing the rotation.

SKILLS ACTIVITY

Investigate different kinds of smart and modern materials and create a table containing the following information.

- Find out more about the properties of some of these materials and their potential uses and applications.
- Find and give details of a product on the market that utilises smart materials.

KNOWLEDGE CHECK

1. Explain how a smart and modern material differs from a natural material. (2)
2. Describe the properties of a smart and modern material and give examples of how these could be used to add functionality to a product. (4)

> **KEY TERMS**
>
> **biomimicry:** an approach to innovation and design that takes inspiration from how nature solves problems. Materials developed using biomimicry include polymers with adhesive properties inspired by a gecko's feet and fabrics with streamlining properties inspired by shark skin.
>
> **memory:** the property by which a material returns to its original shape under specific environmental conditions, such as ambient temperature or heat from an electrical current
>
> **nanoparticles:** nanoparticles are between 1 and 1000 nanometers (10^{-9}) m, and materials where the structure has been engineered at the nanoscale can have unique properties. For example, graphene (comprised of carbon sheets with the thickness of a single atom) has over 200 times the strength of steel, gram for gram. Nanomaterials is the name given to materials manufactured at the molecular level to provide specific properties.
>
> **polymer:** a substance comprised of large molecules made up of simple molecules of the same kind. Plastics such as acrylic and polythene are the most commonly recognised polymers, but polymers exist in nature; for example silk, wool and cellulose.

3.3 Plastics

LEARNING OBJECTIVES
By the end of this section you should:
- ✓ have a working knowledge of the following thermoforming plastics and their properties: nylon, low and high density polyethylene (LDPE and HDPE), polyethylene terephthalate (PET), polyvinyl chloride (PVC), acrylic (PMMA), polystyrene (PS), polypropylene (PP), and acrylonitrile-butadiene-styrene (ABS)
- ✓ have a working knowledge of the following thermosetting plastics and their properties: polyester resin including GRP, melamine formaldehyde (MF), urea formaldehyde (UF), phenol formaldehyde (PF), and epoxy resin.

PLASTIC OR POLYMER?

Plastic is the common term used for materials that are more correctly known as polymers. Polymers are materials formed of molecules in long chains. The term 'plastic' refers to the property of a material to have its shape or form changed. Most polymers are synthetic and derived from petroleum oil, such as nylon, although there are some found naturally, such as silk. There are two types of synthetic polymer: **thermoforming** and **thermosetting**.

Heating/Thermoforming Characteristics for Extruded Acrylic (PMMA)

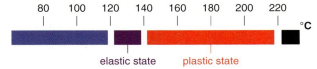

THERMOFORMING PLASTICS

Thermoforming plastics, or thermoplastics, is the term used to describe polymers that are formed when heated, that is, they soften and become malleable. Each polymer will have a temperature band known as its elastic zone where it can be deformed and moulded. Thermoforming plastics can be reheated and reformed; however, repeated working will damage the structure of the material. Thermoforming plastics can be shaped in their **elastic state** using the processes of line bending, vacuum forming, dome blowing or blow moulding, and in their **plastic state** using Rotational Moulding and Injection Moulding. Common thermoforming plastics are shown in the following table.

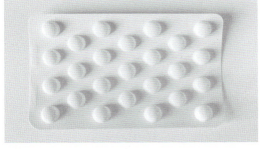

△ Blister packaging is used for pharmaceuticals.

Material	Description	Application
Nylon	Commonly used as a fibre in clothing and textiles, nylon is also a common engineering thermoplastic (TP) available in rod, tube, sheet and powder. Nylon resists abrasion and has low friction characteristics and good chemical resistance.	Mechanical components, such as gears, bushes, bearings and cams, and in fibre form: rope, fabrics, toothbrush bristles, and weather proof coatings.
Low and high density polyethylene (LDPE and HDPE)	The most common TP. Low density polyethylene (LDPE) is less expensive and commonly known as 'polythene' (trade name). High density polyethylene (HDPE) is more expensive and is available in a range of densities. PEs have good resistance to chemicals and some grades float in water.	Injection and blow-moulded bottles and containers, waste bins, packaging films (for example heat shrink), electrical insulation, pallets, toys and bullet-proof vests.
Polyethylene terephthalate (PET)	PET is a hard, stiff, strong and stable material, with low water absorption. It also has good chemical resistance (with the exception of alkalis) and gas barrier properties.	Signs, drinks bottles and bank/smart cards.
Polyvinyl chloride (PVC)	Good resistance to chemicals and solvents. Good tensile strength, reasonably impact resistant and some grades are flexible. Available coloured or clear. PVC is also resistant to water and fire (when burning it releases chlorine atoms that inhibit combustion).	Rain, water and sewage pipes (and fittings), toys, clothing and footwear, adhesive tapes, medical goods and packaging.
Acrylic (PMMA)	Acrylic is hard, with reasonable tensile strength and good resistance to UV light and weather. Available in a range of colours (transparent and opaque) – clear sheets are more transparent than glass. Available in cast and extruded form – both are easily cemented with a solvent.	Bathroom furniture (for example baths), aircraft canopies, headlight lenses, signs, and kitchen products.
Polystyrene (PS)	One of the most commonly used plastics. High Impact Polystyrene (HIPS) is a mixture of polybutadiene and polystyrene, having better impact resistance. Poor resistance to UV light.	Toys, light fittings, computer cases, buttons, car fittings, freezer/refrigerator linings, packaging trays, domestic appliance (for example hairdryers), food packaging.
Polypropylene (PP)	High resistance to chemicals and virtually impervious to water. Highly flexible and resistant to stress.	Chemical containers and tanks, carpets (interior and exterior), fuel cans, toys, marine ropes and underwater bearings, dishwasher - safe food containers, household appliances.
Acrylonitrile-butadiene-styrene (ABS)	Durable, weather and chemical resistant. ABS is rigid with rubber-like characteristics, giving it good impact resistance.	Boat/dinghy hulls, domestic appliance housings (for example telephones) and dashboards, casings for power and garden tools.

THERMOSETTING PLASTICS

Thermosetting plastic materials harden when heat is applied. Typically this is achieved by adding a chemical that causes a reaction, producing heat. A common example is epoxy resin. Theromsetting is a one-way reaction; that is, a thermoset plastic cannot be returned to its original state. Thermosets can be shaped by casting or as a resin encapsulating glassfibre or carbonfibre. They can also be formed by compression moulding. Common thermosetting plastics are shown in the following table.

Material	Description	Application
Melamine formaldehyde (MF)	A clear, hard and chemical resistant thermosetting resin, often known simply as melamine.	Electrical plugs, switches and sockets, gear wheels (laminated with fabric), adhesive to bond plywood layers.
Urea formaldehyde (UF)	High tensile strength and surface hardness. Low water absorption and mould shrinkage.	Buttons, toilet seats, electrical fittings, adhesive to bond pressed wood products, such as MDF.
Phenol formaldehyde (PF)	A synthetic polymer with high hardness, good thermal stability and chemical resistance.	Laboratory worktops, billiard balls, printed circuits boards.
Polyester resin	Stable, low cost and easy to use. Good mechanical, chemical-resistance and electrical properties.	Glass reinforced for boat hulls, car body panels, adhesives, coatings, printed circuits boards.
Epoxy resin	High adhesive strength and mechanical properties, as well as high electrical insulation and good chemical resistance.	Adhesives, sealants, coatings (especially marine protection), moulds/dies for thermoforming plastic parts, printed circuits boards.

SKILLS ACTIVITY

Choose a household product and do the following:

- list the plastic materials that you can identify using the recycling codes
- evaluate the recyclability of the product, including how easily its plastic components can be disassembled and recycled.

> **DESIGN IN ACTION**
>
> A common concern about the use of oil-based plastics is that they tend not to be biodegradable (break down or decompose over time). Sustainable use of materials tries to reduce, reuse or recycle materials. To help with this, there are international symbols to indicate the status of plastic materials. The 'triangle loop'
>
>
>
> symbol (above) indicates whether a material is recyclable. For plastic objects or containers, each element should be marked with a number (resin indication code) indicating the material used. For example, 1 is PET, 2 is HDPE, 3 is PVC, 4 is HDPE, 5 is PP, 6 is PS, 7 is 'other' (various minor plastics) and 9 is ABS.

KNOWLEDGE CHECK

Giving examples of products and specific plastic materials:
1. Explain the advantages of plastics when designing and making everyday products. (3)
2. Explain the disadvantages of plastics when designing and making everyday products. (3)
3. Explain why plastics are commonly used in everyday products. (3)

> **KEY TERMS**
>
> **elastic state:** the state in which a thermoforming polymer is at a temperature that produces a stretchy consistency and in which it can be shaped, but will not retain its form without being held in place. The elastic state is used for instance in line bending, vacuum forming or dome blowing.
>
> **plastic state:** the state in which a thermoforming polymer is at a temperature that produces a consistency that is softer and more malleable than the elastic state. The plastic state is used for instance in injection moulding.
>
> **thermoforming:** plastic materials that can be repeatedly softened by heat and formed into shapes which become hard when cooled
>
> **thermosetting:** plastic materials that can be softened and formed into shapes which become hard when cooled and cannot be softened again

3.4 Wood

> **LEARNING OBJECTIVES**
>
> By the end of this unit you should:
> ✓ be able to state the classifications of natural timbers and understand how natural timbers are classified
> ✓ understand why timber is seasoned and how to care for it
> ✓ understand the process of timber steaming and bending, and have knowledge of adhesives, curing times and strengths
> ✓ have a working knowledge of manufactured boards, and be able to assess advantages and disadvantages of working with them in comparison to natural timbers.

NATURAL TIMBER

Wood that is used in the manufacture of an artefact (for example, furniture, housesa or boats) is referred to as **timber**. The classification of natural timber is by two groups: **hardwoods** and **softwoods**. Importantly, these classifications have nothing to do with how hard the wood actually is, rather these classifications are used to group timbers together that come from similar types of trees. Softwood comes from trees that are coniferous: evergreen trees that bear cones and have needles. Hardwoods come from deciduous trees: trees that shed leaves in autumn. There is one exception to this rule: the holly tree, which is a hardwood and most varieties are also evergreen.

Before timber can be used, trees have to be felled and then processed. Trees tend to grow at different rates and speeds, depending on the species. Generally, hardwoods grow much more slowly than softwoods.

Depending upon how you cut planks from the tree, you can change the properties of the planks. This is because planks produced from different parts will have a different grain structure. Planks that have a wood grain that is closer together are stronger and denser (heavier) than those with a wider spaced wood grain running through it.

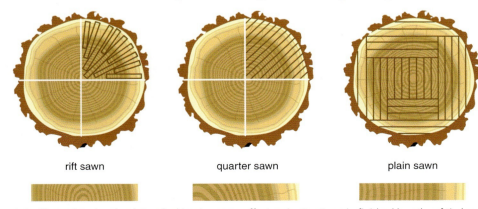

△ An illustration showing how planking trees can affect grain structures in finished lengths of timber.

SEASONING TIMBER

Once cut into planks, timber needs to be seasoned in order to allow it to fully dry out. **Seasoning** is traditionally undertaken by stacking planks of wood and allowing air to circulate between the planks in order to dry them out. This process is now automated and is undertaken on an industrial scale, using a kiln (like the one illustrated below). A kiln used for seasoning timber allows air to be warmed and circulated by fans to speed up the seasoning process. As the warm air circulates around the inside of the kiln, it dries the timber placed inside it. Timber seasoned in this way is referred to as 'kiln dried'.

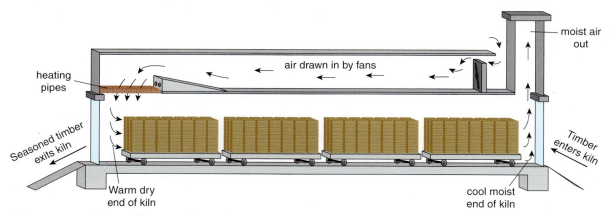

△ An automated kiln used for seasoning timber.

Kiln drying timber to season it has a number of key advantages over naturally seasoned timber. As the moisture content is removed in a controlled fashion, so it is easier to monitor the timber to ensure uniformity throughout its length. It is also a much quicker process than naturally seasoning timber. However, it also has its disadvantages. As it consumes energy as a process, it is more expensive than natural seasoning. A kiln has a maximum capacity due to its physical size, which limits how much timber you can season at anyone time. Due to the forced nature of the seasoning process, sometimes the timber can warp during the process.

It is important to season timber so that it remains strong and true when it is used. If it is not seasoned, it will change its shape (often shrinking) and its properties as it dries out. This is not something that is usually desirable. For example, it would not be ideal for the legs of a chair to bend and change shape after the chair has been assembled and is in use.

To keep seasoned timber in perfect condition, it should ideally be stored somewhere with a controlled environment. This means that it should be kept somewhere that is at a constant temperature and in dry conditions, not somewhere that is damp. It should also be laid flat to avoid it bending and twisting, leading to a condition called **warping**.

SOFTWOOD EXAMPLES

Type of timber	Description	Uses
Scots Pine	Straight-grained but prone to knots, pale in colour, strong yet easy to work with, cheap and readily available	Low-cost furniture, construction work. Simple joinery
Parana Pine	Hard and straight-grained, virtually knot-free, fairly strong, comparatively expensive. Pale yellow in colour with darker brown streaks	Better quality furniture, structural carpentry that is visible such as windows, doors and staircases
Spruce	Creamy-white with small hard knots, not very hardwearing	Indoor furniture including bedrooms and kitchens
Yellow Cedar	Very pale in colour, light in weight yet rigid	Furniture, boat building and veneers
Redwood	Relatively strong, knotty, durable when treated with a suitable coating or treatment, low cost	General woodwork, cupboards, shelves, roofs

HARDWOOD EXAMPLES

Type of timber	Description	Uses
Ash	A light creamy-brown colour, open-grained	Sports equipment, wooden ladders, tools
Beech	White in colour, close-grained, hard and strong, prone to warping	Furniture, toys, tool handles
Elm	Light brown, open grain, tough, resists splitting, durable in water and outdoor settings	Indoor and outdoor furniture
Mahogany	Rich reddish-brown colour, strong and durable, interlocking grain	Good quality furniture
Oak	Light brown colour, strong and tough, open-grained, corrodes steel screws and fittings and reacts with certain adhesives	High quality furniture and interior woodwork

MANUFACTURED BOARDS

There are a number of limitations with natural timber, one of which is the width it is available in - the limiting factor here being the thickness of the tree trunk it is cut from in the first place. Most softwoods come from fast growing trees with a comparatively small diameter trunk. What happens if something larger is required, but the cost of using an expensive hardwood is not an option?

You could join a number of sections together yourself, or you could buy something that has already undergone this process commercially. The term for this is **manufactured boards**. Larger sized manufactured boards tend to be available as a maximum size of 2440 mm × 1220 mm (in imperial measurements, that is 8 ft × 4 ft).

There are a number of different types of manufactured board available, each with different properties. The six most common manufactured boards are shown in the following table.

Blockboard 	Blockboard is manufactured with a central core of softwood strips bonded together with adhesive and covered with a sheet of plywood on either side, and then often a finishing veneer.
Chipboard 	Chipboard is made up of small chips of wood bonded together with resin and compressed to form sheets. It is not as strong as plywood and blockboard, nor does it come in thicker sheet sizes, but it is comparatively cheaper. Chipboard is often used in furniture for use indoors, and it is covered in a plastic coating or veneer for a more aesthetically appealing timber.
Hardboard 	Hardboard is made from pulped wood fibres that are pressurised until the fibres bond together to produce a board that is smooth on one side and rough on the other. It is not as strong as the other boards and it is typically used in non-structural situations, such as the back of cupboards.
Medium Density Fibreboard (MDF) 	MDF is made up from very fine wood dust and resin pressed into a board. This material can be worked, shaped and machined easily and has considerably more strength than hardboard due to the use of a resin as a bonding agent. It is used in many applications, indoors, and it can be easily finished with veneers or paint.

Plywood 	Plywood is made from veneers of timber with the grain of each layer being at right angles to the layers either side of it. The layers are bonded together by resin and pressure. A number of different types of plywood are available, and these are often referred to as grades. They are manufactured differently and designed for different purposes, including: • boil resistant plywood • flexible (flexi) plywood (typically three layers thick with a very thin middle layer) • interior plywood • laser plywood (non-toxic adhesive for use with laser cutters) • marine plywood, which is moisture resistant • weather and boil proof plywood.

There are many advantages to working with manufactured boards as opposed to natural timber:

- they are available in large sizes as well as standard sizes and thicknesses
- boards are designed for specific purposes, so have specific properties (for example, marine plywood is glued together using waterproof adhesive so it can be used in damp and wet environments)
- they often use elements of waste from processing timber, so are environmentally sympathetic (although the adhesive and resin sometimes used is less so)
- manufactured boards are uniform with few imperfections, so when you work with them you are assured that they are not likely to fail due to unseen imperfections
- they do not split like natural timbers do
- they are available in ready finished formats (with **veneers** or plastic coatings pre-applied).

> **REFLECTIVE LOG**
> Identify, and classify, the types of timber and manufactured boards you have used in your own work. Suggest different timbers and manufactured boards that could be substituted for the ones you used. Comment on the relative costs that these changes would bring about. Would it make the product cheaper or more expensive?

TOP TIP

From the outset, consider what you are going to use the timber for, then look for a timber or manufactured board that matches the properties you need. This will give you a range of materials to work with, and you can then make your final decision based on either cost or availability.

SHAPING AND JOINING TIMBER

Timber can be joined in a variety of ways (see unit 3·10). However, many methods involve a wasting process like cutting or machining, or a mechanical fastening like a screw or a nail. What happens if a curved piece of timber is required?

There are two common methods of forming such shapes: steam bending and laminating.

△ A chair produced using steam bending.

Steam bending

Thin layers of timber (veneers) are placed into a steam chamber. Steam is introduced at one end and travels through the chamber, heating the veneer and absorbing into the timber as it does so. As the steam cools, and condenses it turns back into water and it simply drains away.

After a period of time immersed in the steam, the veneers are removed and are malleable and flexible. This change in their properties makes it possible to bend them to a different shape. They are positioned around a former and clamped, or held, into place and left to cool. Upon being unclamped and cooled, they will retain this new shape.

Laminating

This is a process that involves no heat. It involves using a number of thin laminates of timber and bonding them to each other over a former. Unlike steam bending, there is no need to try to get the sections into a steam chamber, and they retain their shape by being bonded to other layers of laminates. Whilst the adhesives used in this process cure and set, the layers of laminates need to be held in place.

Adhesive curing times and strengths vary from product to product. PVA (Polyvinyl Acetate) adhesive is commonly used to bond timber as it is relatively cheap, non-toxic, and easy to work with. PVA is a white liquid, which, when dry after application and exposure to air, becomes transparent. There are many trade names and types. Typically PVA is left to set for a period of 24 hours before removing the clamps, to ensure that it is fully set. Synthetic powdered resins can also be used. Often these are water resistant and they need to be mixed with water before application and use. They also tend to be more expensive than PVA.

TOP TIP

If you are trying to bend and form shapes in timber, consider using either a ratchet strap or the opposing half of the former to hold the laminates in place while they set, or cool while they cure if you are steam bending.

ENVIRONMENTAL CONSIDERATIONS

Trees are a natural resource. They remove harmful gases from the atmosphere, notably carbon dioxide, replacing it with oxygen. As we harvest trees it is essential that we replant them so that the ecological impact is minimised and the cycle can continue for years to come. Some tropical hardwoods; for example, teak and mahogany, take many hundreds of years to grow, so it is essential that we continue to plant new trees in order to take their place. The Forest Stewardship Council (FSC) is a global organisation set up to certify that timber sourced from their members is done so with these basic principals in mind. Timber from an accredited FSC supplier displays their logo, so look out for it when you are buying timber.

SKILLS ACTIVITY

Think about a piece of flat-packed furniture, sketch the item you have thought of and identify where you could use hardwood, softwood and manufactured boards in its construction. Suggest two different types of timber or manufactured board for each of the components you have identified.

KNOWLEDGE CHECK

1. When is wood referred to as timber? (1)
2. What type of trees do softwoods come from? Give three examples and suggest uses for them. (7)
3. What type of trees do hardwoods come from? Give three examples and suggest uses for them. (7)
4. Why are manufactured boards used? (1)
5. State two processes that can be used to bend and shape timber or manufactured boards. (2)
6. Suggest two types of adhesive that can be used to bond timber. (2)
7. How would you know if timbers have come from a sustainable forest? (1)

KEY TERMS

hardwood: wood that comes from broad-leaved trees, such as oak, beech and ash, that lose their leaves in autumn

manufactured board: board such as chipboard or hardboard that is made in specific sizes and with specific properties. It is made by joining or compressing small pieces of wood, often offcuts and waste from timber processing.

seasoning: the process of drying out timber so that it becomes strong and will not change its shape over time

softwood: wood that comes from fast-growing evergreen trees with cones and needles

timber: wood that has been processed and turned into a form that can be easily used, often planks

veneer: a thin layer of a substance (often timber) that is applied as a surface coating to something, often to increase the aesthetic appeal of the underlying material

warping: the distortion or twisting that can occur to timber, often as a result of poor storage, poor seasoning or natural defect

3·5 Composites

> **LEARNING OBJECTIVES**
> By the end of this unit you should:
> ✓ understand the term *composite*
> ✓ be able to name a range of composite materials
> ✓ understand the advantages of using composite materials
> ✓ be aware of the practical applications of a range of composite materials.

WHAT IS A COMPOSITE?

Composite is the term used to describe a material that is made by combining two or more materials. Typically the individual materials have significantly different chemical or physical properties. When combined, each material's structure is retained. This means that the individual materials do not blend, merge or dissolve into each other, and it is quite easy to identify the original properties of each within the new material.

In a composite, the individual materials work together and have enhanced properties, resulting in a new and unique material that may be lighter in weight, have increased rigidity, flexibility or strength.

Most composites are made from just two materials, each performing a different function within the new material. One is known as the **matrix**, which surrounds or binds the other material, which is the fibre, also known as the **reinforcement**.

WHY ARE COMPOSITES USED?

Composites are used because they can be engineered to meet the exact requirements of a specific application. By combining individual materials, each becomes enhanced and typically the advantages of the new material means it is more efficient, stronger and lighter.

EARLY COMPOSITES

Composites are not new, and one of the earliest known examples is combining mud (wet soil or clay) with plant material, such as straw, to make bricks from which walls and buildings could be constructed. In this example the mud is the matrix and the straw acts as the reinforcement.

Concrete, which comprises small stones and gravel with cement and sand, is another early example of a composite material. In many buildings today, concrete is further reinforced by embedding metal wires or rods into the concrete during construction.

△ Steel reinforced concrete

MODERN COMPOSITES

Continuous research is helping to ensure that new innovative composite materials are developed. Within resistant materials there are three main composite groups, as follows:

- Carbon-fibre Reinforced Polymers (CFRP)
- Glass-fibre Reinforced Polymers (GRP)
- Aramid products (Kevlar®)

Glass-fibre Reinforced Polymers (GRP)

Glass-fibre Reinforced Polymers (GRP), also known as fibreglass, is a relatively modern composite material. It is made from strands of glass, which form a flexible matrix or fabric. In the finished product, variation during manufacture in the thickness of layers of glass strands produces different weights and strengths of GRP. Once assembled the strands are set into the desired shape by placing them into a mould. To set the strands in place polyester resin is added and, once the resin has **cured**, the process is repeated. This creates layers of GRP which is lightweight and very strong. In GRP the glass-fibre woven fabric is the reinforcement, and the polyester resin is the matrix. The fine, self-coloured finish of GRP is provided by using a female mould, which has a wax resist applied before laying up the chopped strand and encasing with polyester resin. This method is suited to batch production and will result in the final mouldings having an identical finish.

△ Glass-fibre reinforced polymer components being manufactured.

When sufficient layers have been added, the GRP can be smoothed via sanding and, if desired, a finish can be applied. GRP is a popular composite which is widely used in the manufacture of car bodies, water tanks, swimming pool slides, canoes and boat hulls.

△ A GRP boat hull

Carbon-fibre Reinforced Polymers (CFRP)

Carbon-Fibre Reinforced Polymers (CFRP) is one of the most expensive composite materials, but boasts the best strength-to-weight ratio of any construction material. This composite is made from high-tensile-strength carbon fibres which are woven together and then encased in a polymer resin.

Carbon fibres are resistant to stretching and the result is a rigid material, light in weight yet very strong. CFRP is used extensively within the aerospace industry, it has a high tolerance to heat, and is used to replace traditional, heavier metal-based materials. Other market applications of this versatile composite include marine, automotive, defence, and in the sports and leisure industries.

△ CFRP is used in the manufacture of Formula 1 car bodies.

Kevlar®

Stronger than steel, when woven and manufactured into garments Kevlar® is probably best known as the material used in the manufacture of bullet-proof vests.

Kevlar® is one of a family of materials developed from a synthetic fibre created by chemist Stephanie Kwolek in 1965, while she was developing a new material to make lighter tyres. Formed by combining paraphenylenediamine and terephthaloyl chloride the result was a super stiff, heat resistant polymer, nine times stronger than nylon. When spun, aromatic polyamide (aramid) threads are formed which, when further refined, is woven to create this amazing high-performance, super strong, flexible, lightweight material.

△ Kevlar® reinforced personal body armour.

On its own, Kevlar® is not a composite material, but when layers of the woven fabric are combined with resin, the result is an extremely rigid, lightweight material that has 20 times the strength of steel. A common use for this synthetically engineered material is Personal Protective Equipment (PPE), to protect people working in hazardous jobs.

△ Composites are used in PPE garments.

> **REFLECTIVE LOG**
>
> Consider the materials you have used in your own work. Have you used, or could you consider using, a composite? Explain the advantages a composite material choice would have over other more traditional materials.

TOP TIP

A common mistake that often occurs when learning about composites is made in relation to other types of mixed material. Probably the best example of this is metal. When the base elements are combined, for example when copper is mixed with tin to make bronze, this is not a composite material. This is known as an alloy.

The key difference is that within a composite the different materials do not merge, bind or blend together. They retain their individual properties and often it is visually quite easy to see the separate materials within the new material. A good example would be reinforced concrete, where the metal rods are inserted to increase strength, retain their shape and size, and are clearly visible within the concrete.

SKILLS ACTIVITY

Choose one example of a composite material and complete the following:

- Name each individual material utilised in its construction.
- Describe how the composite is formed, and explain the purpose of each individual material. Which is the matrix and which is the reinforcement?
- Explain the advantages of using this composite over traditional materials, and give examples of its application.

DESIGN IN ACTION

Higher, further, faster! That is the impact of using the composite CFRP as a construction material. CFRP is now used extensively in both professional and recreational sports to make everything from racing car bodies and cycle frames, to tennis rackets, archery bows, skis, snowboards and golf clubs. It is rigid, lightweight and its strength properties have made it an indispensable material in high-performance sport.

△ CFRP is used in the manufacture of sporting prosthetic limbs.

A recent example is in cycling, where lighter CFRP frames and aerodynamic carbon wheels in bicycle construction have meant less exertion, because of the stiffness and lighter weight, and better surface hold on the road or track. However, the use of carbon fibre technology and engineered CFRP composite materials in sport is not without its controversy.

In the 1970s, American amputee Van Phillips designed the very first pair of running blades. Engineered from around 80 layers of carbon fibre, current blades are more sophisticated than Phillips' original but the advantages of using the composite are the same. Each pair can be made specifically to meet the requirements of the individual athlete, taking into account their weight, height and leg length in order to maximise their performance on the track.

Whether you agree or disagree with this use of technology to enhance an athlete's performance in sport, it is amazing to think that it will not be long before our Paralympic athletes are recording faster times on the track than their able-bodied counterparts.

KNOWLEDGE CHECK

1. Give an example of an early composite and name the individual materials used. (3)
2. Using your example from Question 1, state which material is the matrix and which is the reinforcement. (2)
3. Give three benefits of using composite materials. (3)

KEY TERMS

composite: a material that is made from two or more materials which have significantly different physical or chemical properties. When combined they create a single material which is often lighter and stronger than the original individual materials on their own.

cure: the process of allowing a substance, for instance a polymer resin used within a composite, to harden and set

matrix: the material in a composite that performs the binding function

reinforcement: the material in a composite that provides structure and adds strength

3·6 Metals

LEARNING OBJECTIVES
By the end of this unit you should:
- ✓ understand where metals come from and how they are categorised
- ✓ be able to show a working knowledge of a range of metals and provide examples of their uses
- ✓ understand how processes can change the molecular structure of a material, making it more or less suitable for the task it has to perform.

METAL EXTRACTION AND CLASSIFICATION

Metal is found throughout the world. It is mined from the rock that makes up the surface of our planet and comes from the earth in raw form called ore. It is from ore that the metal is extracted and processed. Typically ore is crushed and then heat and chemicals are used to further refine it. Sometimes the ore from which a metal is extracted is evident, other times it is not. Iron comes from iron ore, while aluminium comes from bauxite.

△ Casting iron, following extraction from iron ore.

Pure metals fall into one of two main categories: **ferrous** (those containing iron) and **non-ferrous** (those that do not contain iron). However, pure metals often do not have the exact properties that are required for specific functions, so it is necessary to combine a number of metals to give the desired properties in a process called alloying. Therefore, an **alloy** is a mixture of two or more pure metals. Some examples of alloys and their uses are shown in the following tables of ferrous and non-ferrous metals.

PROPERTIES OF METALS

Metals have many different characteristics. These are described by the various properties they possess:

- **Brittleness**
  something that has no flexibility: when it breaks, it shatters into multiple pieces
- **Ductility**
  the ability to be stretched, bent, deformed and shaped without breaking
- **Hardness**
  resistance to scratching, cutting and wear
- **Elasticity**
  the ability to return to its original shape after it has been deformed

- **Malleability**
  something that can be easily shaped, spread, flattened, hammered and deformed
- **Toughness**
  resistance to breaking, bending or deforming
- **Work hardening**
  When the structure of a metal alters as a result of consistent working. Here the term *working* refers to the carrying out of repeated processes multiple times on the same piece of material, such as repeated hammering of a piece of metal to shape it.

FERROUS METALS

These are metals and alloys that contain iron. As such, all ferrous metals have some common properties which relate to iron: they conduct heat and electricity, and they react with oxygen and corrode (rust) unless they are treated to prevent this. Typically, due to the iron content, ferrous metals also have high melting points (1200°C and greater).

Name	Melting point (°C)	Properties	Uses
Cast iron	1200	Brittle, corrodes by rusting.	Casting and base metal for all steel alloys, manhole covers, car brake discs.
Stainless steel*	1400	Hard, corrosion and wear resistant to a large extent. Typically an alloy of iron with 18% chromium, 8% magnesium and 8% nickel added.	Cutlery, kitchen equipment, sanitary equipment, surgical equipment.
High speed steel (HSS)*	1400	Contains a high content of tungsten, vanadium and chromium alloyed with iron. Very brittle and very resistant to wear.	Drill bits, lathe tools, and milling cutters.
Mild steel*	1600	Tough, ductile, malleable, good tensile strength, poor resistance to corrosion. An alloy of iron combined with 0·15–0·30% carbon.	General purpose engineering and construction material, nuts, bolts, car body panels.
High carbon steel*	1800	Hard and tough but the increase in carbon makes it more brittle. It can be heat-treated (see later) to further enhance its properties.	Cutting tools, ball bearings, hand tools; screwdrivers, hammers, chisels, and saws.

* Indicates that these metals are alloys.

NON-FERROUS METALS

These are metals and metal alloys that do not contain iron.

Name	Melting point (°C)	Properties	Uses
Lead	160	Lead is very soft and very malleable. It has one of the lowest melting points of a metal that is solid at room temperature.	Roofing, construction, casting, lead acid batteries. (At one time it was used in paints and water pipes, but it is toxic and no longer used for these.)
Tin	230	Very ductile and very malleable. It is resistant to corrosion from moisture. Bright silver in appearance, it is often used as a coating on a base metal.	Used as a coating on food cans, beer cans, and tin foil for use in the kitchen.
Zinc	420	By itself, very weak. However, it is highly corrosion-resistant.	Used as a coating on screws, nails, nuts and bolts. It is also used to weatherproof steel in a process called galvanising.
Duraluminium*	630	Very strong, hard, lightweight alloy of aluminium, copper (3%), manganese (1%) and magnesium (1.5%). Soft and workable in normal state.	Widely used in aircraft construction, boat building and car manufacture.
Aluminium	660	Light in colour although it can be polished to a mirror-like appearance. It is very light in weight and the most abundant metal in the earth's surface.	Saucepans, cooking foil, takeaway containers, window frames, ladders, bicycles.
Brass*	925	Is often machined from cast ingots and billets of raw material. It is golden yellow in colour and is typically a mixture of copper (65%) and zinc (35%).	Used for decorative metal work such as handles, candlesticks, ornaments, pins on electrical plugs, and musical instruments.
Bronze*	950	Deep brown/yellow in colour, usually cast into shape. It darkens with exposure to the elements.	Bearings, artistic and sculptural pieces, hand bells.
Copper	1080	A very ductile and malleable metal. It is often red/brown in colour and it goes green when it corrodes, forming a substance called verdigris. It is an excellent conductor of heat and electricity.	Used for plumbing, cookware, electrical fittings and roof coverings.

* Indicates that these metals are alloys.

Other low-melt casting alloys exist and they are the materials that are often used in schools. They are often based on aluminium with the addition of copper (3%) and silicon (3–4%). Pewter has often been associated with this kind of use, but low-melt casting alloys are proving to be excellent substitues as they are more stable, readily available, comparatively cheaper, and more predictable in use. These properties exist in low-melt casting alloys because they are manufactured to provide superior consistency during use than pewter. They are not as heavily reliant on tin as a base metal (pewter is typically made with 85–99% tin), which is more costly and more scarce than aluminium used in low-melt casting alloys.

CHANGING THE STRUCTURE AND PROPERTIES OF A METAL
Work hardening

Work hardening occurs when a non-ferrous metal is continually bent, hit or shaped over a period of time. This is often referred to as *working* the metal. Work hardening can be both desirable and undesirable.

If a piece of structural aluminium on a car or an aeroplane is work hardened then it could split and fail, giving rise to a terrible accident. This will occur as the work hardened section is more rigid and less likely to bend and deform in use, so when it fails it is often suddenly and with catastrophic consequences. On the other hand, if you were to manufacture something from a piece of copper it may be desirable to heat it so it becomes really malleable and easy to shape. Once it is in the desired shape it will need to be hardened to retain the new shape it is in. A process called planishing is often used to do this. Planishing involves intentionally work hardening the copper by striking the shaped artefact many times with a planishing hammer, thereby hardening it with each strike of the hammer. This process intentionally work hardens the copper.

Annealing

Annealing is a process that is often undertaken after a piece of material is work hardened, and it is in a state where it can no longer be worked as required. When a metal is annealed, it is heated to a specific temperature and then cooled – it does not involve working with the material in a hot state. Heating and cooling changes the molecular structure of the material, and these changes result in the metal being softened. This means that a metal that has been annealed can be cut and shaped more easily than if it had not been.

During the annealing process, the metal needs to be heated to a specific temperature. As you heat a piece of metal it changes colour, and these changes in colour occur at specific temperatures. Consequently, it is often easier to look at the colour of a piece of heated metal and use this as an indication of its temperature, rather than try to measure it with a thermometer. If we consider the alloy *mild steel*, this metal needs to be heated to a red colour and allowed to cool in order to anneal it.

However, metals with lower melting points, such as aluminium, will melt if heated for too long. Aluminium can be annealed but great care must be taken while heating. If this is being done in a small-scale workshop (not an industrial setting), then the flame should be held at a distance away from the aluminium so that it gets heated uniformly.

TOP TIP

A useful trick when annealing aluminium is to rub a bar of soap on to the surface of the metal and then heat it. Once the soap turns black, immediately remove the heat from the aluminium and allow it to slowly cool. It should now be successfully annealed and very easy to work with.

It is also important to ensure that the flame being used to anneal the aluminium is not being held too close to the metal, or it will not heat uniformly.

Hardening and tempering of tool steel

If tool steel is hardened it becomes too brittle to be of any use, so a secondary process called *tempering* needs to be used in order to make metal useable. Consider, for example, the blade of a screwdriver. It needs to be very strong and hard, yet not brittle or it will not be able to withstand the turning force (torque) generated in turning a screw. To harden the screwdriver blade, the tool steel is formed into the desired shape and then it is heated until it is at a temperature of around 900°C. This is easily observed as the metal will turn red with the heat. Once it reaches this temperature it is plunged into clean, cold water.

To remove the brittleness but retain the strength, the tool steel must now be tempered. In order to do this, the metal must first be thoroughly cleaned. Once cleaned, it is very gradually reheated. As it is heated, it changes colour and heating continues until a blue coloured line is seen on the screwdriver blade, which signifies that the metal is now at the correct temperature. It then needs to be cooled, either by quenching in cold water again, or oil can be used. The cooling can be done more gradually by placing the hot metal onto a larger piece of metal (like an anvil) so that the heat dissipates away from the work. The following chart indicates the temperature and colour for tempering tool steel.

colour	temperature (°C)	example
pale yellow	230	scriber
dark yellow	250	twist drill bit
brown	260	plane blades
purple	270	knives
dark purple	280	saws
blue	300	screwdrivers

△ Temperature and colour chart for tempering tool steel.

Case hardening of mild steel

Case hardening is quicker and less complex, but less precise than the hardening and tempering process used on tool steel. This process hardens the surface of the metal only, unlike hardening and tempering, which uniformly hardens the material throughout. Typically this process only hardens the outer 0·05 mm. This process also has the advantage of leaving the core unhardened, so it retains greater malleability and flexibility, allowing the material to still be easily worked. Should you need to increase the depth of the hardened surface, the process can be repeated, building up a deeper layer of material.

Initially the mild steel is heated to a red colour. Following this it is removed from the heat and plunged into a case hardening compound that is high in carbon, as it is the addition of this element that hardens the steel. Once the metal has been immersed in the case hardening compound, it is returned to the heat and again its temperature raised until it glows red. At this stage it is removed from the heat and plunged into clean, cold water.

TOP TIP

When selecting a metal, consider the following: does it need to be resistant to corrosion? How strong does it need to be? What type of surface finish needs to be applied? Once you have done this, you can then determine if you need to undertake a hardening (or softening) process in order to work with it and make it durable enough for the task you require from it.

SKILLS ACTIVITY

Identify four uses of metal in tools that you know and have used. Determine a suitable metal for each tool and work out if they are likely to be ferrous or non-ferrous (could they be alloys?). Looking at the tools again, work out if they need to be hardened and, if so, which processes would be suitable?

Again, thinking about the tools you have identified, note any specific metal features, such as a saw blade, then list the properties that the metals must have in order to fulfill their function.

KNOWLEDGE CHECK

1. Explain how to determine if a metal is ferrous or non-ferrous. (2)
2. Describe in simple terms what an alloy is. (2)
3. Put these metals in order according to their melting points: aluminium, bronze, cast iron, copper, duraluminium, lead, mild steel, stainless steel, tin, zinc. (10)
4. What is the purpose of annealing metals? (1)
5. When and why would a metal be tempered? (2)
6. Typically, how thick is the hardened layer on mild steel that has been case hardened? (1)
7. Identify four steps in the process of case hardening, in sequence. Draw a picture of each step, labelling the parts in each drawing. (8)

REFLECTIVE LOG Explain why you think it is necessary to recycle metals, thinking specifically about alloys. Why are these harder to recycle than metals in their pure form?

 KEY TERMS

alloy: a metallic material that is a mixture of two metals

ferrous: a metal that contains iron and which will rust unless protected, for example by being painted

non-ferrous: describes a metal that does not contain iron and so will not rust

3·7 Preparation of materials

LEARNING OBJECTIVES

By the end of this unit you should:
- ✓ be able to demonstrate knowledge of available market forms, types and sizes
- ✓ understand different methods of cutting
- ✓ understand the use of datum surfaces, lines and edges and know how to produce them
- ✓ be able to explain the preparation needed for machine processes and safe methods of securing materials to work surfaces.

WHAT IS MARKET FORM?

Market form describes manufactured standard shapes and sizes of material, wood or metal available to buy commercially.

When planning a project, knowledge of market forms, shapes and sizes of material has several advantages. Effective use of this knowledge helps to ensure that project costs are kept to a minimum, particularly when working in metal. By using pre-cut material, time is saved and subsequently labour costs are reduced. Using market forms is also better for the environment because it means there is less waste material. This illustration shows industry standard forms, shapes and sizes for metal and wood.

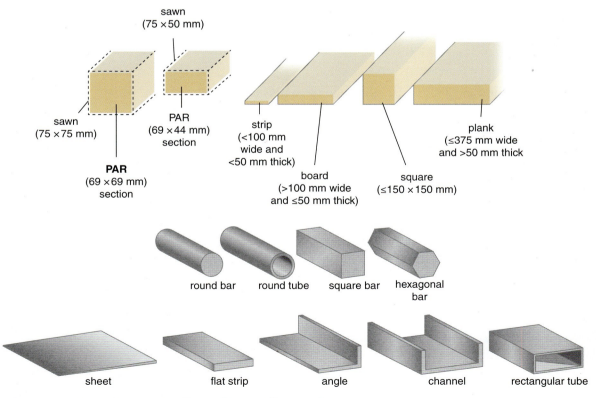

△ Examples of standard sectional profiles available in different materials.

METHODS OF CUTTING

In order to select the most appropriate tool you must first consider not only the type of material, but its length, thickness, shape and form. Of course, you also need to consider the finished length, thickness, shape and form you need the material to be.

Cutting tool	Materials it is suitable for use with	Uses
Hacksaw	All metals	Cuts straight lines
Guillotine	Sheet metal	Cuts straight lines
Tenon saw	Wood	Cuts precision tenon joints
Cross-cut saw	Wood	Cuts across the grain, producing a straight edge
Panel saw	Suitable for heavy duty work, cutting all types of wood, and sheets of plastic	Cuts straight lines

Portable power tools

Portable power tools are especially useful when undertaking work outside of the workshop environment. It is important to carefully follow the manufacturer's instructions when using power tools and selecting the type, shape and size of material to be cut. Portable circular saws cut straight lines, whereas a jigsaw can be used, within the manufacturer's guidance, to create more intricate shapes.

△ A portable circular saw being used to trim timber sections to length.

CREATING A DATUM SURFACE

A **datum** is the flat face or straight edge of material from which all measurements should be taken. Smoothing a newly cut edge will produce a datum surface, but the method of creating the datum varies depending upon the material. To produce a datum on timber a wood plane is used, whereas on plastic or metal a flat hand file is required.

To check the surface is level, a steel rule or straight edge may be used, but to accurately produce datum lines on metals it is best to use a surface plate in conjunction with a scribing block or calipers.

Some tasks will require two datum surfaces, which should be **perpendicular** to each other. In order to ensure that another surface is perpendicular a try square is used. When this is done on timber the surfaces are known as a face side and face edge.

SKILLS ACTIVITY

You are preparing a piece of timber in advance of marking out. You need to produce a datum surface. List the tools and equipment you will need. Explain, step by step, the stages you would take.

SAFE METHODS OF SECURING MATERIAL

Holding devices such as cramps, vices and jigs can be used to ensure material is held securely in position during the processes of cutting, shaping or forming the material. The type of device used will depend on the process employed, and also by the shape, size and type of material.

△ Assorted clamps are often used to hold materials securely in position.

Incorrectly securing material can lead to undesired consequences. The material may become loose during machining, which could result in damage not only to the material itself but also to the equipment. In extreme instances the material may snap or shatter causing sharp pieces of material to break off potentially causing injury to people nearby.

Material can be secured safely by clamping the material directly to the work surface, work table or sturdy bench using appropriate mechanisms including jigs, cramps and wood and metal vices.

During machining, secure the material firmly to the faceplate of a lathe, or between the centres on a lathe, in accordance with the manufacturer's recommendation. It is important to ensure that the jaws of the chucks, used to hold either tools or work, are tight and material is appropriate in size and weight for the process being undertaken.

TOP TIP

When mistakes are made in cutting, shaping, bending and forming material this usually occurs as a result of poor preparation.

No matter how accurate your measuring, setting and marking out may be, if the edge of the material from which you take your initial measurement is not flat, perpendicular or square, joints will be low in quality and the final product may not fit together properly.

DESIGN IN ACTION

In the building and construction industry, knowing how to accurately record and calculate measurements is crucial. In preparing a quote for a client, the ability to calculate the exact amount of material required is essential. A good working knowledge of market form, shape and size not only helps ensure you purchase the correct amount of material, but also that you order material which is pre-cut to an exact size, shape or form, which will save time and also reduces wastage.

Inaccurate calculations lead not only to inefficiency but may result in a quote which is too high, and the client deciding to use another company. In contrast, a quote that is too low will result in little or no profit, and may even end up in financial loss for the company.

KNOWLEDGE CHECK

1. List two things you should wear to protect yourself when preparing or cutting materials. (2)
2. Give two methods of holding material securely while cutting and shaping. (2)
3. Give two potential consequences associated with poor preparation. (2)

REFLECTIVE LOG

How have available market forms, and types and shapes of material, influenced design decisions in your work?

KEY TERMS

datum: a point used as a reference point for all measurements when marking out material

market form: standard forms and sizes of commercially manufactured material

PAR (Planed All Round): the market form of wood with planed edges. The actual size of PAR timber is smaller than the stated size owing to planing. A 50 mm × 25 mm PAR will therefore measure about 45 mm × 20 mm.

perpendicular: lines or surfaces that are at right angles (90°) to each other

3·8 Setting, measuring, marking out and testing

LEARNING OBJECTIVES
By the end of this unit you should:
- ✓ understand the importance of accurate measurement
- ✓ understand setting, marking out and testing key terms
- ✓ be aware of the following tools and equipment: rule, pencil, marking knife, marker pen, scriber, try square, bevel, mitre square, centre square, dot/centre punch, dividers, inside/outside/odd-leg calipers, template, marking/cutting/mortise gauge
- ✓ be able to accurately produce datum lines
- ✓ understand why testing is important.

WHAT IS SETTING AND MARKING OUT?

In its simplest terms, setting and **marking out** describe the first stages of the manufacturing process where measurements- which can be in the form of lines or shapes- are transferred onto material prior to, and in preparation for, cutting, shaping, bending or forming. Accuracy is essential in order to ensure that each individual part of the product, when assembled, fits together properly.

WHY IS ACCURACY IN MEASUREMENT IMPORTANT?

When purchasing material for a project, if the measurements are inaccurate this could mean you get less than you have paid for and you would not have sufficient material to work with. Inaccurate measuring in your own work could mean that pieces do not fit together, which could reduce both the function and the aesethic appeal.

> **DESIGN IN ACTION**
>
> Everyone makes mistakes! There have been lots of mathematical errors that have led to catastrophe over the years, but probably one of the most memorable measurement mishaps occurred in 1999, and resulted in NASA losing its $125 million Mars probe.
>
> The Climate Orbiter was designed to relay environmental data about Mars back to Earth, but engineers failed to convert British imperial measurements into metric units. Two hundred and eight-six days into its mission, the simple error was discovered when the probe fired its engines to push itself into orbit. The mistake caused the satellite to veer 60 km off course, pushing it beyond Mars and out into space.

MEASURING AND MARKING OUT

Accurately measuring, setting and marking out material is critical to help ensure a high quality outcome. To help ensure accuracy, setting, measuring and marking out should always be done in a clean and well lit area. The tools, equipment and methods needed to ensure precision will differ depending upon the material and intended processes of manufacture. Before beginning, all tools and equipment should be checked to ensure they are sound and in good working order.

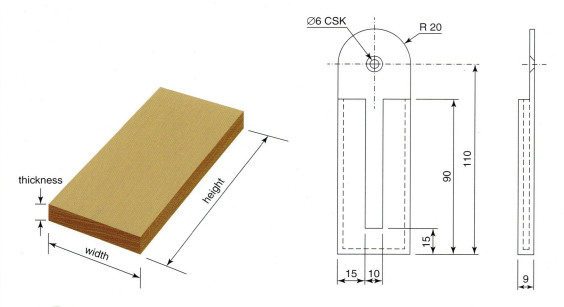

>
> **REFLECTIVE LOG**
>
> Using an example from your own work, write about one occasion when you measured and marked out material.
>
> - What preparation did you do?
> - Which tools and equipment did you use?
> - If you were marking out a different material, would you have used the same tools and equipment?

>
> **KEY TERMS**
>
> **marking out:** the process of transferring measurements in the form of lines or shapes onto a material in advance of cutting, shaping, bending or forming

SELECTING THE CORRECT TOOLS AND EQUIPMENT

The table on the next page shows the correct methods, tools and techniques to use in order to ensure accuracy when you set, mark out and measure different materials.

Tools for measuring and marking out

Type of mark	Correct tool/equipment		
	Timber	Plastic	Metal
A sharp, straight line (any angle)	Sharp pencil, marking knife or sliding bevel	Permanent marker pen	Straight steel edge and scriber
A parallel line	Marking gauge	Odd-leg calipers	Odd-leg calipers
A perpendicular/ 90° angle	Try square	Try square	Engineer's square
A 45° angle	Mitre square/ combination square	Mitre square/ combination square	Mitre square/ combination square
The centre of a hole	Sharp pencil and a centre punch	Permanent marker pen	Dot/centre punch
The centre of a round piece of material	Centre square	Centre square	Centre square
Arches, curves and circles	A pair of compasses	Dividers	Dividers
An irregular shape	Template	Template	Template
A mortise and tenon joint	A mortise gauge for both mortise and tenon	-	-

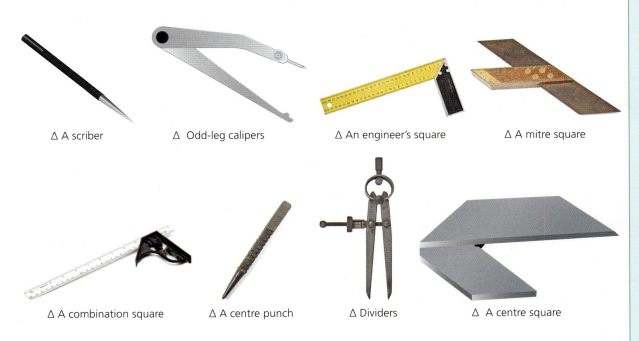

△ A scriber △ Odd-leg calipers △ An engineer's square △ A mitre square

△ A combination square △ A centre punch △ Dividers △ A centre square

Precision instruments for measurement

To ensure high accuracy and absolute precision, where measurements of fractions of a millimetre are required, it is necessary to use a micrometre or calipers. These are explained below:

- **Inside and outside calipers:** Calipers are used to measure the inside or outside of material.
- **Micrometre:** Sometimes also known as a micrometre screw gauge, this is a precision device used to measure depth, length or thickness to a high degree of accuracy.
- **Vernier Gauge:** The Vernier Gauge is a manual caliper used to measure internal and external distances with precision and accuracy. It is less easy to read than the digital caliper but does not require any power source.
- **Digital vernier caliper:** A precision instrument used to accurately and easily measure internal and external distances. The LCD display can be set to measure either imperial or metric units.

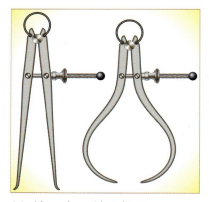
△ Inside and outside calipers

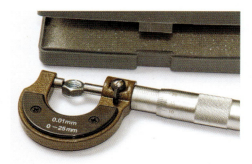
△ A micrometer

△ A vernier gauge

△ A digital vernier caliper

TOP TIP

Incorrect equipment, such as using a pencil to mark out instead of a scriber or marking knife, can vary the thickness of the line, and may lead to inaccuracy in cutting.

Avoid using a non-permanent marker pen on plastic, which may result in marks that become blurred or rub off.

Try to make as few marks as possible. Numerous marks may make it unclear which line you need to cut along.

CREATING A DATUM USING A SURFACE PLATE AND CALIPERS

All measurements should be taken from the datum line and it is good practice to use a surface plate as this helps to ensure accuracy. A surface plate is a large, solid, flat plate, typically made from a hard material such as granite or steel. Surface plates are used as the support material during the process of setting and marking out. Once the datum is created it can be checked using the surface plate.

TOP TIP

Most common mistakes made when measuring and marking out occur when you become distracted, are in a hurry, or are tired. Each of these factors will cause you to lose concentration, which will increase the risk of making a mistake.

Probably the most common error, however, occurs when cumulative measurements are taken. This happens when, instead of taking each measurement from the datum line, once an initial measurement is taken, the next is taken from the last, and so on. To avoid this simple error, and improve accuracy, always start each measurement from the datum.

SKILLS ACTIVITY

In preparation for cutting you have been asked to mark out a piece of mild steel. List the tools and equipment you would use. Explain your choices. Describe each stage of the process you would do.

KNOWLEDGE CHECK

1. Explain why accurate measurement is important. (2)
2. Give two methods of setting and marking out wood, and list the tools required. (2)
3. Give two methods of setting and marking out metal, and list the tools required. (2)
4. Give two potential consquences of inaccurate measurement. (2)

3.9 Shaping

LEARNING OBJECTIVES

By the end of this unit you should:
- ✓ understand common wastage and addition processes
- ✓ understand common deforming and reforming processes.

PRINCIPLES OF SHAPING MATERIALS

Each category of materials has a different set of tools and equipment that are used to shape them. However, the principles of **wastage**, **addition**, **deforming** and **reforming** apply across the range of **resistant materials**. In school workshops, the most common tools used to shape materials by wastage are saws, files and other abrasives (for example, glass paper). Similarly, tools and equipment used to shape by addition are screws, nails, adhesives, bolts and rivets. Deforming processes, such as laminating woods or vacuum forming, are used to create shapes while retaining (or enhancing) structural integrity. Reforming processes, such as casting metals or injection moulding plastics, are used to form complex shapes without the need to employ wastage techniques.

△ Laminated wood is often used for furniture.

△ Vacuum formed plastic is often used for packaging.

△ Cast metal is often used in construction.

△ Injection moulded plastic is often used in machinery components.

SHAPING BY WASTAGE AND ADDITION

Designers, technologists and engineers need to be aware of a range of tools and methods for cutting and removing material, as well as methods for joining materials to produce desired shapes, forms or contours. This knowledge provides a wider range of options when designing and making high quality products and prototypes. The tools for shaping woods, metals and plastics (for example, hand, portable, power, machine, or CAD) are distinctly different, due to the properties of materials.

Hand tools

Process	Woods	Metals	Plastics
Cutting straight lines	Tenon saw	Hacksaw or junior hacksaw	Scoring tool or hacksaw
Cutting curved lines	Coping saw	Piercing saw	Coping saw or piercing saw
Shaping edges	Planes, spokeshaves, rasps	Files	Files
Smoothing surfaces	Abrasive papers	Abrasive papers and cloths, polish	Abrasive papers and cloths, polish
Hollowing	Chisels, hand router and gouges	Cold chisels	N/A
Drilling holes	Wheel brace and bit	Hand drill and twist bit	Hand drill and twist bit

△ A tenon saw can be used for cutting straight lines when working with wood.

△ A hacksaw can be used on metals and plastics for cutting straight lines.

△ A coping saw can be used for cutting curved lines when working with wood or plastics.

△ A piercing saw can cut curved lines into metal and plastics.

△ Rasps can be used to shape the edges of wood.

△ Planes can shape the edges of different types of wood.

△ Chisels are useful for hollowing when working with wood.

△ A hand router can be used for hollowing when working with wood.

△ Gouges are often used for hollowing woods.

△ A wheel brace and bit drills holes into wood.

△ Hand drill and twist bit can drill holes into both metals and plastics.

△ Abrasive papers can be used for smoothing surfaces when working with wood, metal or plastic.

Machine tools

Process	Woods	Metals	Plastics
Cutting straight lines	Circular saw	Guillotine, metal cutting bandsaw	Circular saw, bandsaw
Cutting curved lines	Bandsaw	Bandsaw	Bandsaw
Shaping edges	Router, disk/belt sander, lathe	Off-hand grinder, centre lathe	Disk/belt sander
Smoothing surfaces	Planner thicknesser, belt sander	Polishing machines, belt sander	Polishing machines
Hollowing	Router	Milling machining	Milling machining, router
Drilling holes	Pillar/bench drill	Pillar/bench drill	Pillar/bench drill

△ A circular saw cuts straight lines into woods and plastics.

△ A bandsaw is a useful tool for cutting curved lines when working with woods, metals and plastics.

△ A belt sander can shape the edges of woods and plastics as well as smooth the surfaces of metals and woods.

△ A pillar/bench drill can drill holes into woods, metals and plastics.

△ A lathe can shape the edges of wood. A centre lathe is needed to shape the edges of metals.

△ A milling machine is used for hollowing metals and plastics.

Computer-aided manufacture (CAM) typically uses wastage and includes:

- laser and plasma cutters (wastage)
- 3- and 4-axis milling machines (wastage)
- 3- and 4-axis routing machines (wastage)
- Computer Numeric Control (CNC) lathes machines (wastage)
- 3D printing, including Fusion Deposition Modelling (FDM) machines (addition).

Drilling

Drilling is a useful method for making a hole all the way (through hole) or partially (blind hole) through a material. It can also be used to remove larger amounts of material before, for example, using a chisel to remove a square section of wood from a tenon joint (see unit 3·10). Hole boring or drilling can be achieved by hand or machine, depending on the material. Different elements of a hole used for joining two materials include:

- **pilot hole:** smaller than the screw being used, so that the thread cuts in and holds (usually in the bottom piece so that it is pulled against the top piece)
- **clearance hole:** larger than the thread and shank of a screw (usually in the top piece), so that the screw does not grip or 'bite' into the sides, allowing the two pieces to pull together
- **tapping:** usually used with metals, this is where a screw-thread is added to the inside of a pilot hole, using a tool called a 'tap', so that it accepts a machine screw. A thread can be added to the outside of a rod using a tool called a die
- **countersunk hole:** a conical shaped indentation at the top of a hole to accept a countersink headed screw, with the screw head being 'flush' (level) with the surface, usually with a 90 degree inclusive angle
- **counterbored hole:** a larger diameter recess at the top of a hole to accept a cheese or fillister headed screw, with the screw head being recessed below the surface.

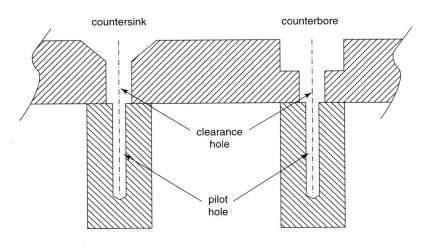

SHAPING BY DEFORMING AND REFORMING

Deforming and reforming processes are used to modify the shape of an object from its standard form (for example, planks, bars, sheets) without removing (wastage) or combining (addition) material. In the classroom you may use some of the processes described below. Deforming and reforming is used in industry to form complex shapes and gain strength from an object being formed as a single piece.

Deforming processes

Process	Materials	Description	Applications
Steam bending	Hardwoods, such as beech and oak	Strips of hardwood are heated in a steam chamber until the wood is pliable enough to easily bend around a former to create a specific shape.	Wooden furniture
Lamination	Hardwoods, such as birch and ash	Thin sheets (veneers) of hardwoods are layered with strong adhesive and pressed in a former until the adhesive cures.	Wooden furniture, decorative roof beams
Press forming	Sheet materials (woods, metals and plastics)	Sheet materials are pressed between a male and female (or saddle and yolk) former, until they cool, to take on a new shape. Some materials need to be heated to become 'plastic' (malleable) prior to forming, or in the case of metals they can be annealed.	Metal vehicle panels
Vacuum forming	Thermoforming plastics	Plastic sheets are heated to their elastic state, then a former is pressed into the sheet and the air below is evacuated. The external pressure presses the plastic onto the former.	Food packaging/trays
Blow moulding	Thermoforming plastics	Tubes of plastic are heated and fed through a mould. The air is blown into the tube, forcing it to expand out and form against the mould.	Plastic milk bottles

> **REFLECTIVE LOG** Referring to the tables in this unit, make a note of the tools and equipment that you have used. Identify them in a Design & Technology workshop, and locate tools and equipment that you have yet to use. Investigate, and ask your teacher (a) how they are used, (b) what they do and (c) what materials they are used with.

Reforming processes

Process	Materials	Description	Applications
Injection moulding	Thermoforming plastics	Plastic beads are heated to their plastic state and injected into a mould to form complex shapes.	Toys, consumer goods, such as mobile phone cases
Extrusion	Thermoforming plastics and metals	The material is heated to its plastic state and pushed through a 'die' to create components or parts with a fixed cross-sectional profile.	Pipes, tubes, beams and rods
Sand casting	Metals	Non-ferrous metal (for example iron, brass aluminium) is melted and poured into a mould made of a special sand.	Engine blocks, gas and water valves
Die casting	Non-ferrous metals	The metal is melted and forced into a hardened tool steel mould (or die).	Toys, brackets, engine parts

SKILLS ACTIVITY

Think about your most recent design-and-make assignment. Which tools and equipment did you use? Sketch your product and label/annotate where you used wastage, addition, deforming and/or reforming.

KNOWLEDGE CHECK

1. Explain the differences between wastage, addition, deforming and reforming. (8)
2. Choose one example each for wood, metal, plastics and composite, and describe in detail a process using each of the four principles of shaping (wastage, addition, deforming and reforming). (4)

KEY TERMS

addition: the process of shaping materials by combining or joining them, for instance using screws, nails, nuts and bolts, and adhesives. Joining methods can be classified as temporary or permanent, with particular methods being used with different materials.

deforming: the process of subjecting a material to a stress that changes its shape. Methods include bending and the use of jigs and formers. Typically heat is applied to materials to bring them into their 'elastic' zone, where they can be deformed.

reforming: the process of changing the shape of a material, typically by melting and pouring or injecting the molten material into a mould, for example when injection moulding plastics or sand casting non-ferrous metals

resistant material: a material that requires force in order to be worked or shaped

wastage: the process of cutting away material to leave the desired shape, for instance using saws, files and abrasives

3·10 Joining and assembly

LEARNING OBJECTIVES
By the end of this unit you should:
- ✓ be able to define both permanent and temporary methods of joining materials and provide examples of both
- ✓ be able to use holding devices, jigs and formers to aid construction
- ✓ understand the process of soldering, brazing, welding, riveting/pop riveting
- ✓ understand methods of constructing artefacts using timber
- ✓ recognise the use of knock-down fittings for use with manufactured boards
- ✓ understand methods of carcase, stool and frame construction including different types of joints
- ✓ describe how screws, nails, nuts and bolts are specified and understand how different sizes are described and defined
- ✓ give examples of different types of adhesives and list their uses and limitations.

TEMPORARY JOINING

Temporary joining of materials can be achieved using components that have a screw thread. Typical examples include: screws, nuts and bolts, set screws and knock-down fittings (commonly found in flat-packed furniture).

△ There are lots of different types of nuts, bolts and set screws.

PERMANENT JOINING

Permanent joining of materials can be achieved using, as the name suggests, a process that is not designed to be easily reversed. Processes such as welding, brazing and soldering involve using heat to create a bond between metals. All forms of riveting are also considered to be permanent joining processes. Nailing pieces of timber together is a method of permanent joining, along with any type of adhesive application between parts.

HOLDING DEVICES: JIGS AND FORMERS

Jigs and formers are used to hold pieces of material together in order for them to be joined. By using this technique it is possible to exactly replicate the method of joining many times. It also allows small and delicate pieces of materials to be joined more easily, as they can be held in place while the joining method is applied. This is especially the case when adhesives are used as a joining method.

BRAZING

Brazing is a process often used to join two different metals together. The process involves melting a third metal (often brass) over, and into, the two pieces to be joined. This only works if the two pieces to be joined together have a higher melting point than the brass being used to join them. The pieces are heated together using a brazing torch, the brass is introduced into the joint, and the heat remains there until the brass melts. Once the brass is molten the heat is removed and the brass cools and solidifies, bonding the two pieces being joined together.

SOLDERING

Soldering is a type of brazing which uses a lower-melting-point alloy as its filler material. The consequence of this is that soldering is often weaker than brazing, so it needs to be selected accordingly with this in mind. Soft soldering is used to join copper, brass or tinplate. It is also the way electronic components are joined. Soft solder typically melts at around 200°C. Once heated, solder will 'flow' along the cleaned components bonding them together once it solidifies. Hard solder melts at around 625°C, and is used for stronger joints.

WELDING

Welding differs from soldering as the metals to be joined are heated to extreme temperatures, causing the components to melt. Once they have cooled and solidified, they are then joined together. A filler material is used during the process to keep the molten material clean and free from oxidation during cooling. Oxidation is the name given to the process whereby oxygen interacts with the surface of a metal and it causes that metal to corrode. Each metal reacts differently to oxygen; for example, when iron oxidises it produces a red, brown layer which we call 'rust'.

There are different types of welding processes. Some involve using electricity (electric-arc) to generate a spark in order to melt the metals to be joined, and others involve using a flammable gas mixed with oxygen (oxy-acetylene) to achieve the same outcome. Different methods are more suited to different types of metal due to the differences in melting points of metals. As the temperatures generated in welding are very high, it is very important that people joining materials by welding wear appropriate eye protection.

△ Weld fillers joining two pieces of mild steel together.

△ Electric-arc welding

RIVETING

A rivet is a permanent method of joining, where a small piece of metal is inserted into a hole in each of the components being joined. It is then deformed to an extent that it cannot be removed.

△ Riveted structural steel

There are a number of different rivet profiles. Countersunk rivets allow the head of the rivet to sit in the material being riveted. This has the advantage of making the top surface of the material and the rivet flush (level). For some applications, this is essential. For example, the wing of an aeroplane needs to be smooth so that airflow over it is not disturbed.

Often riveted material is seen with big round dome rivets protruding out from the surface of the metal. These are used where the dome shape on the surface of the material being joined is not a problem. On an industrial scale these are applied using machines; on a small scale they are often applied using a 'snap and set', seen below. These are tools that are used in pairs to ensure that the shape of the rivet is maintained.

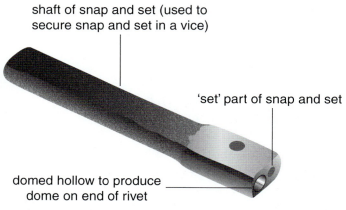

shaft of snap and set (used to secure snap and set in a vice)

'set' part of snap and set

domed hollow to produce dome on end of rivet

△ A rivet snap and set

The following seven steps indicate how to create a riveted joint, accompanied by a diagram over the page.

1. Using the domed hollow in a snap and set, a rivet head is placed in it with the shank of the rivet pointing upwards. The two pieces of material to be joined are then slid over this.
2. Another snap and set is placed on top of the exposed rivet shank and it is hammered down. This forces the pieces together and makes sure there is no gap between them.
3. The rivet shank is cut to length using a pair of pinchers, pliers or bolt cutters.
4. Using the flat surface on a ball pein hammer, the exposed end of the rivet is deformed.
5. The hammer is turned over and, with the ball pein end, it is roughly shaped into a dome.

6. A second snap and set is then placed over the slightly deformed end of the exposed rivet and it is hammered down. The domed hollow in the snap and set shapes the end of the rivet into a smooth, regular dome.

7. A complete riveted joint is created.

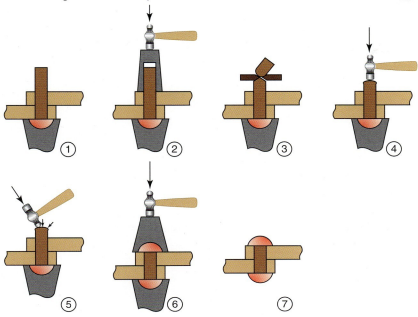

△ There are seven key stages to creating a riveted joint using a ball pein hammer and a 'snap and set'.

Another type of rivet is the pop rivet. This involves a more modern process where a two-piece rivet and a special tool, called a rivet gun, is used to fit them. These rivets have a softer part, the rivet head, and a harder part, the rivet shank. The rivet is located into a hole going through both pieces of the material to be joined. The shank is then put into the rivet gun, and is drawn through the softer rivet head and splayed outwards. This has the advantage that the process is completed from one side and can be used in situations where the rivet cannot be supported at the back.

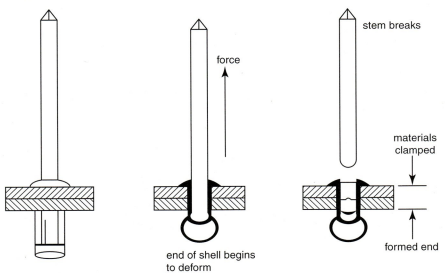

△ How a pop rivet works

JOINING WOOD

Wood joints

There are many different types of wood joint. Some require little or no adhesive to hold them together (for example, dovetail joints). Others increase the surface area at the site of the joint so allowing more adhesive to be used to increase the strength (for example, comb/finger joint). There are also joining mechanisms, which are designed to be used in manufactured boards. These provide a positive mechanical lock joining the two pieces together (for example, a cam lock).

Consider a piece of carcase furniture (furniture with a box-like construction, such as chest of drawers or bookcases). There are a number of broad types of joints that can be seen in the work, notably corner joints and joints where pieces of timber intersect each other.

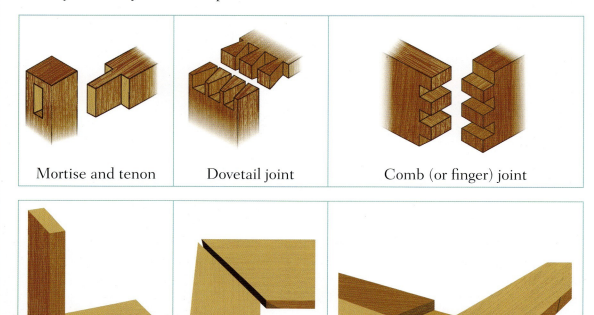

△ Common corner joints used in wood

Knock-down (KD) fittings

Wood joints have been around for thousands of years. However, as technology has advanced, so have the methods available for joining wood. The rise in the popularity of flat-packed furniture has led to a need for different methods of joining components that can be easily carried out in the home by a non-technical expert with the minimum amount of equipment.

When products are shipped with knock-down fittings, it is beneficial to try and minimise the number of different fittings used. This is cheaper for the manufacturer as they do not have to carry large stocks of different components, or use a lot of different machinery in order to machine out the holes and slots into which the fittings locate. Also, from an end user perspective, there is less chance of using the wrong fitting if a lower number of fitting types are used.

The simplest form of knock-down fitting is a corner block. These are usually manufactured in plastic. However, a square section piece of timber can be used if it is drilled in the appropriate places. Both of these are installed in the same way.

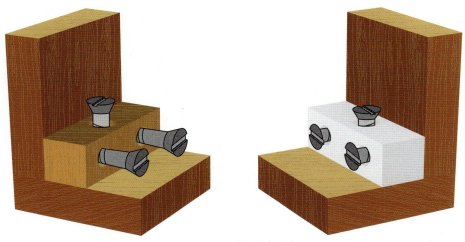

△ A wooden corner block (left); a plastic corner block (right)

To separate out the parts, once joined, it is necessary to unscrew one of the pieces from the timber to which it is connected. Gaining access to do this is not easy. There is an alternative, which involves using two blocks joined together by a bolt and structurally supported by additional steel pins.

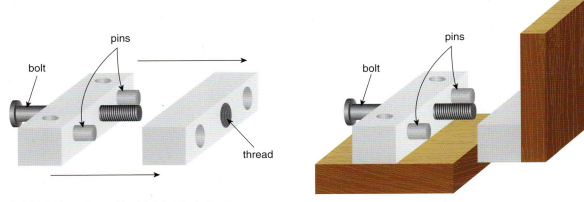

△ A two-piece corner block joining technique

Scan fittings are used to pull together pieces very securely. They are so effective that the join they create can be considered permanent in some cases. A steel barrel fitting is inserted into one piece of the work. In the middle of it is a threaded hole into which a screw is inserted through the second piece of work and tightened.

One of the more favoured types of KD fitting, particularly in manufactured boards and sheet material, is a cam lock. This requires drilling a shallow hole, into which the cam is located, and another hole in the adjoining piece, into which a specially manufactured bolt type piece is located. These two pieces of the fitting are brought together and then the cam is turned, locking the pieces in position.

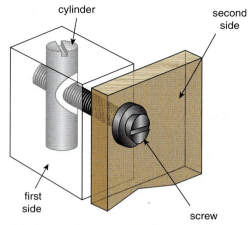

△ A diagram showing a scan fitting

PRE-MANUFACTURED COMPONENTS

There are a number of pre-manufactured components that can also be used to join materials together. These include screws, nails, nuts, bolts, hinges and catches. Some will join only certain materials, and others will join a range of dissimilar materials, like metal to wood, for example.

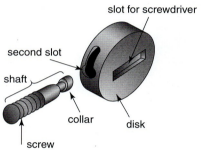

△ This cam lock is in an open position.

Threaded fasteners

There are a large range of screws and bolts with different profile head shapes. Some of these are designed to lie flush with the surface of the material they are in (they are countersunk). Others stand out from the surface. Screws have a pointed tip and are designed to locate and drive into a surface so they are not suitable for use on metal. Often it is necessary to drill a pilot hole into the material before driving in the screw. This makes it locate in the right place and avoids splitting the material. Bolts have a flat profile at the end opposing the head. The thread is uniform all the way down the bolt as it is designed to accommodate a nut of some form.

Wood screw Screws with a smooth shank below the head, they have a tapered point and are used in wood.	
Machine screw Screws with threads up to the head. They do not taper and are designed to be used with nuts or threaded holes.	

As well as a range of threaded fasteners, there are a range of different types of screwdrivers. These vary by having different shaped end profiles to accommodate different shaped holes and slots in the head of the screw. In addition to slotted screwdrivers, there are Pozidriv and Phillips screwdrivers, both of which are cross-headed screwdrivers. There are spanners and sockets for use with bolts. There are unusual and unique head shapes and fittings to prevent casual users without specialist tools from removing them. These are mostly used for security.

Woodscrews come in sizes called gauges. This is a measurement of the diameter of the metal from which the screw is manufactured. In addition to this measurement they also come in a range of set lengths, usually specified in millimetres.

Diameter		Common Lengths	
Gauge	Metric (mm)	Imperial (inches)	Metric (mm)
2	2·0	$\frac{1}{4} - \frac{1}{2}$	6·5–13
3	2·5	$\frac{3}{8} - 1$	10–25
4	3·0	$\frac{1}{4} - \frac{11}{2}$	6·5–38
6	3·5	$\frac{3}{8} - \frac{21}{2}$	10–65
8	4·0	$\frac{3}{8} - \frac{31}{2}$	10–90
10	5·0	$\frac{1}{2} - 4$	13–100
12	5·5	$\frac{3}{4} - 4$	19–125
14	6·5	$\frac{11}{4} - 6$	32–150

Nuts and bolts are much simpler. They are described by their length and their diameter. If they are metric (in millimetres) then the diameter has the letter M before it. For example, a bolt that is M5 × 45 mm has a thread diameter of 5 mm and a length of 45 mm.

Nuts also come in a range of sizes, shapes and profiles, each being designed to undertake different jobs, two of which are highlighted below.

Hex Six-sided nut in common usage.	
Nylon insert lock A nut with a nylon inner piece to keep it in place (often referred to as a nyloc nut).	

Nails

Nails are used in timber, or to fix things to timber. They tend to be used more in softwood than hardwood, as they can bend and distort in hardwoods due to the dense nature of hardwoods. Nails come in sizes specified by their length in millimetres.

Oval wire nail A long nail used in constructing artefacts from timber. Due to its oval shape, it does not tend to split the timber into which it is being hammered. The longer it is, the harder it becomes to drive deeper into timber. The head lies flush with the surface of the wood into which it is being hammered.	
Panel pin By far the most common method of joining timber with nails in schools, these are small and light and often used in conjunction with an adhesive. They are designed to lie flush with the surface of the material into which they are being secured.	

Hinges

Hinges also come in a variety of different shapes, materials and sizes. They are used to connect two pieces of material that need to move. Hinges can be mounted onto wood and screwed into place, or mounted onto metal and secured with nuts and bolts. They can even be welded into place.

Hinges are mounted between the two pieces to be joined, or adjacent to the pieces if they are decorative and to be seen.

△ Types of hinges

△ This hinge is mounted in a rebated slot so it is virtually hidden.

△ Surface mounted hinges, are decorative, so are always visible on the product.

Catches, locks and latches

These are used to secure lids to boxes and close doors on cabinets. They come in different shapes, sizes and materials. It is important to determine what the exact function of the catch is in order to ensure the correct one is selected: is it merely to hold something in place, open or closed? Alternatively, is it to secure something needing a key, or combination to gain access? It is also important to consider the thickness of the material being joined by the catch. Can the catch be placed into the material, or must it be placed inside, or outside?

Decorative swing latch These can be very simple and plain, or decorative and ornate. They are quick to fix and easy to open and sit outside of the product.	
Mortise chest lock These provide a secure locking mechanism between two pieces of material, and they require a key to operate them.	

ADHESIVES

Adhesives are used to join materials by gluing them together. There are many different types of adhesive. Some are set by exposure to the air (they dry out), some have a spirit-based element that evaporates allowing the adhesive to harden, and others work by mixing different chemicals and forcing a chemical reaction to occur, allowing the adhesive to set. As a result of this, some adhesives take longer to set than others. It is also important to consider health and safety here. As spirit-based adhesives dry, the fumes produced during the drying process will often be flammable and detrimental to your health if inhaled. This is one of the reasons it is very important to use adhesives in a well-ventilated room.

The following table lists the most common types of adhesives and their main uses.

Adhesive	Used for...
Acrylic cement	Joining of acrylic and some other types of plastics. The adhesive 'melts' the surface of the plastic parts being joined and it fuses.
Contact adhesive	Joining polysytrene and fabrics. It can also be used for fixing layers, or coatings, of plastic to a wooden base.
Epoxy resin	Joining metals and plastics. It is waterproof and comes in different parts that need to be mixed by the end user.
PVA (Polyvinyl Acetate)	Contruction and assembly of timber products. Some PVA adhesives are water-resistant.
Synthetic resin	Joining timber where PVA is not suitable. It is water-resistant and it needs to be mixed prior to application.

CARCASE, STOOL AND FRAME CONSTRUCTION

All of the techniques and processes described in the Joining Wood section can be combined to manufacture a vast range of products. These products can be assembled with permanent techniques to ensure that they will last, or with temporary fastenings so that the product can be altered or disassembled as required. Sometimes a variety of techniques can be combined for a better result, for example, using adhesives to secure wood joints. Carcases and frames form significant sub-components in the construction of most items of furniture – undoubtedly they will form parts of the products you are likely to design and make. In these cases, it is important to choose and produce joints that will join the timber together early in the manufacturing process. However, it is a good idea not to permanently secure these joints until later in the build so that shelves, hinges and back panels can be fitted.

Stools and chairs require considered planning; often the legs are fixed underneath the section of the seat you sit on. As such, you do not want the top of the legs going through the seat. Mortise joints are often used for square legs and blind holes (holes that do not go all the way through) are often used for circular legs. As the joints between the seat and the legs of a stool or chair are put under the most stress, they must be securely fastened.

TOP TIPS

Make sure you are clear about what you require the materials you are joining together to do. For example, are you making a product that needs to open and close? Does the material you are joining need to resist a force (tension, compression or shearing)? Also consider whether it needs to be permanent or temporary. This will help you choose the appropriate joining method for the best results.

Regardless of the joining method you are using, from simplest to the most complex, it is very important to prepare the pieces being joined properly. Make sure that they align correctly and that they are able to take the fixing you intend to use.

SKILLS ACTIVITY

Design a piece of flat-packed furniture with a hinged section (a door or a lid). Specify which fixings you would use to assemble the piece you have designed and then, using the internet or an appropriate catalogue, determine the total cost of the fixtures and fittings you have specified.

What else could you use to reduce the cost of the product you have designed?

KNOWLEDGE CHECK

1. Describe the differences between permanent and temporary fastenings. (2)
2. What is the principal difference between welding and brazing? (2)
3. List three different types of corner joints that can be made from wood. Suggest two alternative KD fittings that could be used in their place. (5)
4. What does the gauge of a screw represent? (1)

KEY TERMS

permanent joining: a join that is not required to be separated and cannot be taken apart

temporary joining: a join designed so that it can be easily taken apart

3·11 Finishes

LEARNING OBJECTIVES
By the end of this unit you should:
- ✓ understand what a finish is
- ✓ understand why finishes are important
- ✓ understand which type of finish to use
- ✓ understand how finishes are applied, both for one-off and industrial production methods.

THE IMPORTANCE OF FINISHES

Finish is the term used to describe the process of completing the manufacture, or the addition of decoration to enhance the visual appeal, of an article or product. There are two reasons why we apply finishes to materials:

- **Protection**: To protect the material the product is made from. For example, so that it can withstand or become resistant to heat (flame or fire) or water (moisture or liquid), in order to prevent corrosion, staining, rust, mould and fungus growth. Alternatively, a finish may be applied to make the material stronger, or last longer.
- **Aesthetics**: To enhance the material or product's visual appearance, so that it is more attractive and looks better.

△ An example of protective exterior finish.

△ An example of polished cast finish.

△ An example of self-finishing material.

TYPES OF FINISH

There are many different types of finish that can be applied to a product or material. The type of finish needed will be determined by the material and by the function required of the finish itself.

Grouped by material (wood, metal and plastic), the following charts present a range of suitable finishes for each material, providing examples and a brief description of the application process needed.

Metal

Finish	Material	How it is applied	Why it is applied	Example in action
Galvanising	Iron, steel	Via a chemical process which coats the base material with zinc	The zinc provides a good rust and corrosion-resistant finish	Hinges and locks, nails, nuts and screws, car bodies
Anodising	Typically aluminium, but also titanium and magnesium	Via electrolysis, using acid and electrical current to build protective layers onto the base material	To increase resistance to corrosion; also an opportunity to add colour	Flashlights, smartphones, cookware and costume jewellery
Electroplating	Typically iron and steel	Chemically, using electricity, to add layers of other metal (typically nickel or chromium) over the top of the base layer	The additional layer helps to prevent rust and corrosion, and the finish can be highly polished	Chrome plating of motorbike wheels or engine parts
Paint	Any metal	Spray or brush	Prevent corrosion or enhance the appearance	Metal gates, post boxes
Oil/grease	Any metal	Sprayed, rubbed or smeared onto exposed surface areas	Applied as a protective layer to prevent corrosion	The metal structure underneath static caravans
Polish	Any metal	Machine buffing or by hand using a cloth	To add shine to the material	Copper, bronze or brass ornaments, jewellery
Plastic dip coating	Any metal	Usually the metallic object is dipped into a tank of thermoplastic coating, removed, and left to cool	To enhance ergonomic features (make things easier or safer to hold), or make weather resistant	Plastic-coated handles
Powder coating	Any metal	Powder is applied via an electrostatic spray gun. The product is then heated, and the powder melts	To add a hard, protective coating	Washing machines and fridge freezers

Wood

Finish	Material	How it is applied	Why it is applied	Examples in action
Wax polish	Any wood	Buffing or by hand with a cloth	Aesthetic appeal, wax adds a dull sheen, to bring out the grain of the timber	Interior doors, furniture
French polish	Usually hard woods like mahogany	Several very thin layers are applied, usually by hand with a cloth	French polish, made from shellac, gives a high-gloss finish	Expensive furniture
Varnish	All wood	Spray, brush or roller in thin coats	Protection from the elements, also aids aesthetic appeal	Boat decking
Wood stain	All wood	Spray or brush	To apply colour, or to bring out the wood grain	Interior or exterior furniture
Exterior wood stain	Used on both hard and soft woods	Applied with a brush or spray	Used as a preservative to protect the wood from the elements	Sheds and garden furniture, trellis, fencing panels
Oil (teak, Danish, vegetable or linseed)	Any wood	Often applied with a cloth, although it can be sprayed onto larger surfaces for an even coat	This finish penetrates into the wood itself, so it is often longer lasting than surface finishes. It can also enhance the aesthetic appeal of a material	Traditionally linseed oil is used on cricket bats. Vegetable oil can be used on cooking utensils
Paint (oil or water-based)	Any wood	Brush or spray	To add a protective coating that improves visual appeal	Internal doors, door frames and skirting boards
Cellulose sealant	Any wood, but particularly good for porous surfaces such as MDF	Brush or spray	To seal the surface, usually prior to a further finish being applied	Internal furniture

Plastic

Finish	Material	How it is applied	Why it is applied	Example in action
Buffing compound/polish	Injection moulded and shaped plastics	Buffing wheel or hand-polished with a soft cloth	To remove any rough or sharp edges, remove static dust, and to add extra shine for enhanced visual appeal	Children's toys

> **TOP TIP**
>
> A common mistake made in the application of a finish is the lack of thorough **preparation**.
>
> Preparation is essential to help ensure the finish applied will be effective. To ensure you get the best results, make sure the material has been prepared in accordance with the manufacturer's recommendations. This usually means ensuring that the surface is dry and free from grease, mould, mildew, rust or dust.
>
> It often takes time to ensure a really high quality outcome, so be patient. If applying multiple coats of a paint, varnish or stain, apply a series of thin, even coats. Allow each coat to fully dry before applying the next.

CHOOSING AN APPROPRIATE FINISH

Choosing the most suitable finish is very important. To determine the most appropriate type of finish there are two key considerations:

- **The material:** Before you can select an appropriate finish you need to know which type of material it will be applied to. Use the table in this section to support your choice, as well as reading the manufacturer's instructions to ensure an appropriate finish is selected. Selecting an incompatible finish may damage the material or make it ineffective.
- **Function:** In selecting a finish you must understand why you are applying it. Is the finish required for aesthetics or visual appeal? Is it for internal or external use? You need to know how the product will be used and select a finish that will effectively enhance its function.

SELF-FINISHING MATERIALS

New specialist materials are being developed constantly. These plastic-based materials have been engineered specifically to have a 'built-in-finish'. For example, rather than spraying it with a **smart finish** after manufacture in order to make it waterproof, Gortex® is a **self-finished** fabric that has been engineered to be water-resistant.

A common use for these new synthetic materials is Personal Protective Equipment (PPE), to protect

△ A firefighter's PPE clothing made from Nomex®.

people working in hazardous and dangerous jobs. Another example of a self-finished material is Nomex®, which has been engineered to be resistant to heat.

HEALTH AND SAFETY

Before using any finish it is very important to read the manufacturer's instructions carefully. Preparation is not just about the material, it also means thinking about yourself and the environment you are working in. Some finishes may be solvent-based and consequently may be flammable, hazardous or toxic, so it is vital that you follow the manufacturer's recommendations and take precautions to protect yourself. You may need to wear gloves, a face mask or overalls. You may also need to consider the environment in which you are working: is it well ventilated? Do you need to protect the work surface? See unit 1·5 for more on health and safety.

SKILLS ACTIVITY

Choose an industrially manufactured product that is in the room where you are working. List all the materials used in its construction. For each material, list the type of finish used and explain its function. Why has it been applied? Describe the process of application for each finish you have listed.

KNOWLEDGE CHECK

1. Give four reasons why a finish is applied. (4)
2. Name a finish that can be applied to wood, and state how and why it is applied. (3)
3. Give an industrial method of finishing metal. Give the type of metal and a reason why the finish is applied. (3)
4. Name one example of a self-finishing material and give one advantage. (2)

REFLECTIVE LOG Write about a finish that you have used in your own work.

How did you decide which finish to apply? What preparation did you undertake? How was the finished applied?

If your product was manufactured on a larger scale through industrial production, would the same finish be appropriate? Would the finish be applied in the same way?

KEY TERMS

exterior: suitable for outside or outdoor use

interior: suitable for inside or indoor use only

preparation: the process of making the surface of the material ready before applying a finish

self-finishing: a material that does not require a finish

smart finish: a coating that is applied to the surface of a material to enhance its performance, for example waterproofing or flame retardant spray

special finish: a protective coating that is applied to the surface of a material

surface finish: a coating that is applied to the surface of a material in order to enhance its appearance

This section introduces a range of systems and control terms, concepts and ideas in a Design & Technology context. Systems feature widely in the technological world in which we live. They range from simple systems, such as air-conditioning systems used to cool houses, to complex computer systems used to control space craft, satellites and aircraft. Using a systems approach is useful in understanding how an overall system might operate in practice.

Considering a design in systems terms is also useful in the early stages of the design process. Breaking a design into smaller parts can allow a designer to work systematically through a design task rather than tackling the whole design at once, which can be daunting. A common feature of many Design & Technology systems is that they are controlled by electrical and electronic circuitry. This section covers some basic mechanical and electronic principles that will help you design and make a range of systems and control projects. It also highlights the intimate link between electrical and mechanical systems, which is a central feature of many modern systems and control applications.

SECTION CONTENTS

4·1 Systems

4·2 Structures
- 4·2·1 Basic concepts
- 4·2·2 Types of frame structure members
- 4·2·3 Strengthening frame structures
- 4·2·4 Nature of structural members
- 4·2·5 Applied loads and reactions
- 4·2·6 Moments
- 4·2·7 Materials
- 4·2·8 Testing `EXTENDED`
- 4·2·9 Joints in structures
- 4·2·10 Forces

4·3 Mechanisms
- 4·3·1 Basic concepts
- 4·3·2 Conversion of motion
- 4·3·3 Transmission of motion
- 4·3·4 Energy `EXTENDED`
- 4·3·5 Bearings and lubrication

4·4 Electronics
- 4·4·1 Basic concepts
- 4·4·2 Circuit building techniques
- 4·4·3 Switches
- 4·4·4 Resistors
- 4·4·5 Transistors
- 4·4·6 Diodes
- 4·4·7 Transducers
- 4·4·8 Capacitors
- 4·4·9 Time delay circuits
- 4·4·10 Logic gates and operational amplifiers

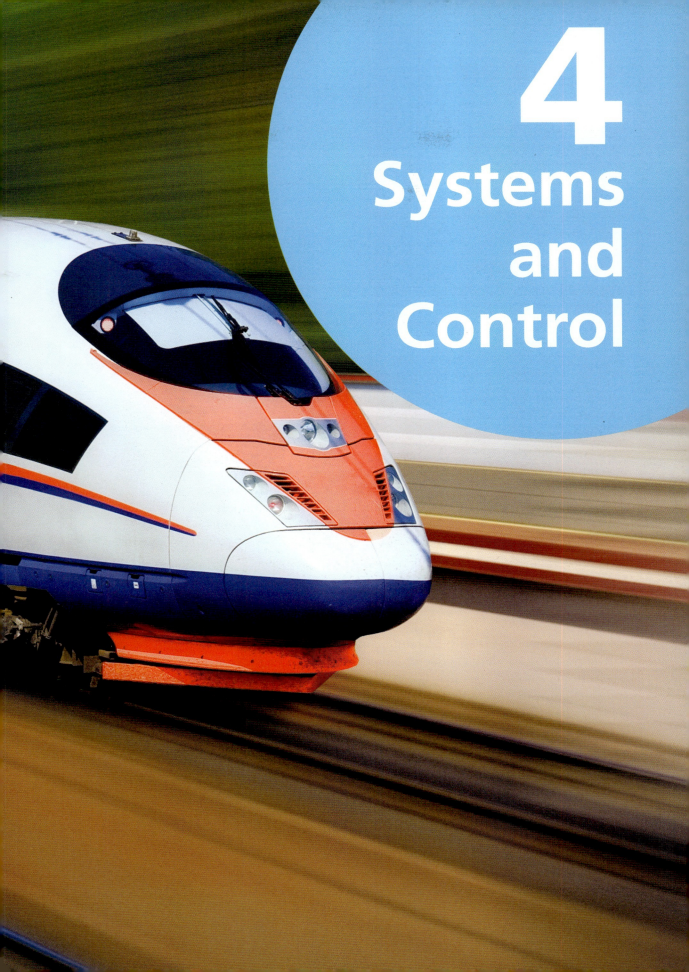

4 Systems and Control

4·1 Systems

LEARNING OBJECTIVES
By the end of this unit you should:
- ✓ be able to describe systems using block diagrams
- ✓ understand that each system block has an input, an output and describes an operation
- ✓ be able to draw a system diagram for a simple system.

WHAT IS A SYSTEM?

A **system** is a set of parts that work together to achieve an output. For example, mechanisms are systems which typically consist of motors, gears, levers, cams, and belts and pulleys. Mechanisms are found in a wide range of products including washing machines, cars, bicycles, sewing machines, lifts and automatic doors.

△ A bicycle is a good example of a mechanism.

Structures can be regarded as systems designed to support an object, load or set of forces; for example, buildings, cranes and bridges.

Many systems are control systems. A control system is a set of equipment used to manage and regulate parts of a system. For example, a thermostat is part of a system that can be used to control and set the output temperature of an air conditioning system. Most modern day control systems are electronic and are used to control and regulate a wide range of machines, products and equipment.

This section covers three systems and control areas:
- structures
- mechanisms
- electronics.

For many applications the three areas are integrated and not considered as individual items. Modern cars, for example, have mechanisms controlled by electronic circuits. Cars may also have computers for satellite navigation and calculating fuel consumption. The body of the car is also a type of shell structure which provides the inside space for passengers.

System diagrams (also called block diagrams) are a common way of describing systems. Each part or subsystem is represented by a single block. Each block has an input, describes an operation and has an output.

Each block can be regarded as a 'black box'. A 'black box' approach views the system solely in terms of its inputs and outputs without needing a detailed understanding of how each part of the system actually works. The details of each block can be developed at a later stage.

When a system has more than one stage then the output of each stage becomes the input for the next stage.

SKILLS ACTIVITY

Draw a block diagram for each of the following household products:

- a microwave oven
- a simple air conditioning system
- a kettle
- a washing machine.

For each product, identify the type of input and output and explain what each stage does.

CLOSED- AND OPEN-LOOP SYSTEMS

An **open-loop system** has no **feedback** to control the system. The accuracy of the system depends on how well the system has been set up before it is operated.

Closed-loop systems use feedback provided by sensors which can sense output values. Sensors provide information which is fed back to the input to modify the input values if required.

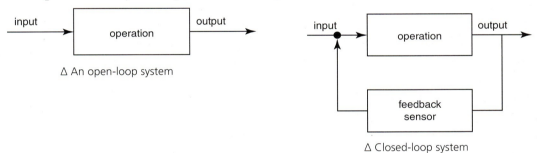

△ An open-loop system

△ Closed-loop system

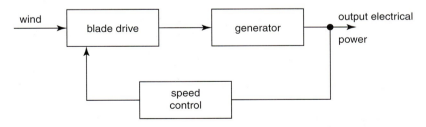

△ A system diagram for a wind turbine

TOP TIP

Whenever you draw system diagrams it is important that you consider how the whole system can be conveniently broken down into smaller parts or stages. Remember that each system block will have an input and output, and that these relate to what each part of the system is trying to achieve.

SKILLS ACTIVITY

Identify three products or appliances in your school. Decide how each could be broken down into a system. Make notes of how each part might link together and sketch a systems diagram for each product or appliance chosen.

KNOWLEDGE CHECK

1. What is the difference between an open- and closed-loop system? (3)
2. How can a framework structure be described as a system? (1)
3. What is meant by an integrated system? (2)
4. How might sensors be used in (i) a burglar alarm system (ii) an automatic door in a supermarket? (2)

KEY TERMS

closed-loop system: a system which incorporates feedback to control and manage the system

feedback: part of a closed loop system which monitors the output and sends a signal back to the input. This signal may cause an adjustment (if necessary), in order to achieve the target output level.

open-loop system: a system which is controlled without the benefit of feedback

system: a set of parts or processes which work together to achieve an output or perform a function

system diagram: a diagram that shows the main parts or functions of a system, usually in the form of a block diagram where they are represented by blocks joined with lines

4·2 Structures

4·2·1 Basic Concepts

> **LEARNING OBJECTIVES**
> By the end of this unit you should:
> ✓ understand that structures can be classified as mass, frame or shell structures
> ✓ be able to give examples of mass, frame and shell structures
> ✓ understand that the principal of levers can be incorporated into a structure
> ✓ understand that frame structures are designed to span gaps and support loads
> ✓ be able to identify structures which occur in nature.

WHAT IS A STRUCTURE?

Structures can be seen everywhere in the world in which we live. For example, they are used for buildings, bridges and transport applications as well as for smaller applications such as food packaging and casings for mobile phones. Some structures, such as sports stadiums, are complex and have many parts. Others are much simpler, for example, egg boxes. However, in each case, the shape and construction of the structure has been designed for a specific purpose.

△ The Colosseum in Rome is a classic example of a mass structure, relying on its weight for stability.

Structures can be grouped as:

- mass structures
- frame structures
- shell structures.

MASS STRUCTURES

Mass structures rely on the weight of material for their strength and support. They are typically built using **reinforced concrete**, stone, brick or soil. For example, in ancient Rome buildings such as the Colosseum were built using stone, and the walls, columns and roofs relied on the weight of the building materials for stability and strength.

River and reservoir dams are examples of much larger mass structures. Dams tend to be built using large quantities of soil, brick, stone or reinforced concrete. These materials help make the overall mass of the structure strong enough to withstand the extremely large forces exerted by the volume of water.

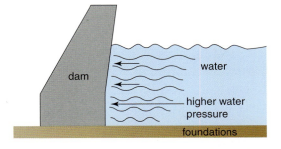

△ Dams are designed to withstand large forces from water pressure. A wider base has more mass.

FRAME STRUCTURES

Frame structures are made by connecting individual parts (called members). Common examples are chairs, bridges, pylons, cranes and roof trusses. The parts are connected so that the overall structure can withstand the various forces acting on it. Each part would not be strong enough to withstand the forces on its own.

△ Steel frame structures form the basis of many modern buildings.

Many buildings use a framework structure to provide a 'skeleton' for the building. These often consist of a welded steel framework onto which glass, metal and other 'cladding' material panels are attached. The panels do not support the main structural forces, but are there to enclose the building and provide protection from the weather. The outer skin materials can also be used to enhance the aesthetics of the building. Frame structures of this type tend to be sturdy and can be built extremely high, for example, skyscrapers. They can also span large gaps and take large loads.

SKILLS ACTIVITY

Select examples of two different bridge-frame structures. Make a sketch of each structure and note down the purpose of each part.

SHELL STRUCTURES

Shell structures are often made from thin materials such as sheet steel. Shell structures create space and protect things inside the structure. A tent is a simple example of a shell structure. Other examples include car bodies, ships' hulls and the fuselages of aircraft. Shell structures can be strengthened by bending and corrugating the sheet material, which can look like the diagrams below.

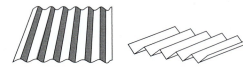

△ The body of a car is a good example of a shell structure.

An example of this is a paper punch, which is made from a number of shell structures joined together. This example also shows that levers can be used as part of a structure or machine to make a job easier to carry out.

When the punch is operated, a second class lever is used to reduce the effort required to punch through a stack of paper. This is a combination of structures and mechanisms. Other examples include a car body that is made from both frame and shell structures and relies on mechanisms for its performance.

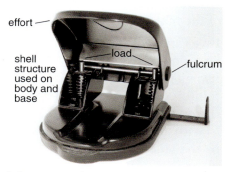

△ A paper punch

△ A carbon fibre shell structure

△ A frame structure

Bicycles are another form of transport where frame structures are vital for the main component, but shell structures, such as the saddle or carbon fibre wheels are parts of the complete machine. In the pictures above, the rear wheels, frames and riders' helmets are made from carbon fibre: a composite material formed into a shell structure.

The frame uses triangulation for strength; the cranks are levers which turn the chain ring and chain. The brakes and gears are operated by lever mechanisms. This is a good example of different types of systems and different materials operating in harmony. The front wheel in the left-hand photo is a carbon fibre shell structure. In the right-hand photo, the front wheel is a frame structure that uses spokes held in tension to keep the wheel rigid.

NATURAL STRUCTURES

Natural structures give useful insights into how man-made structures work and can provide inspiration for structural design. For example, the Olympic Stadium in Beijing is based on a bird's nest, and the Taipei 101 Tower in Taiwan was inspired by the slender shape of bamboo.

The human skeleton is a good example of a natural structure. Its main function is to provide strength and support for the human body. The human skeleton has joints that allow movement of the body in certain directions. The backbone, for example, allows bending and some rotary movement of the upper body. The feet also provide support when the person is standing, walking or running.

Modern manufacturing robots, such as those used to weld car parts, have 'structures', which perform similar functions to the human skeleton. For example, they have a structure designed to make them stable and steady, and have joints which allow them to move in particular directions.

△ The Olympic Stadium in Beijing is based on the natural structure of a bird's nest.

△ A robotic arm, often used in factory work, mimics the functions of the human skeleton.

SKILLS ACTIVITY

Select an outdoor area you know well and identify five natural structures in that area. What is the purpose of each of the natural structures?

TOP TIP

When you are in the process of designing a system or product it is useful to consider natural structures to help you think about what form your design might take.

KNOWLEDGE CHECK

1. Give two examples of each of the following: mass structures, frame structures, shell structures. (3)
2. Explain in your own words what is meant by a mass structure and a frame structure. (2)

KEY TERMS

frame structure: a structure that is constructed using a 'skeleton' of individual parts

mass structure: a structure that is constructed by building or piling up materials such as stone, brick, concrete or soil

natural structure: a structure that occurs in nature and which is created by animals, birds, insects, fauna and flora and geological features

reinforced concrete: a composite material that uses steel to strengthen concrete for structural use

shell structure: an external structure that has space inside

4·2·2 Types of frame structure members

LEARNING OBJECTIVES

By the end of the unit you should:
- ✓ understand that frame structures can be constructed using ties, struts, beams and columns
- ✓ understand that ties resist pulling (tension) forces
- ✓ understand that struts resist compressive (squeezing) forces
- ✓ understand that beams are designed to resist bending
- ✓ understand that columns resist compressive forces.

A frame structure is a system of parts connected to support the loads acting on it. Each part is called a **member**. In practice, the structure should be strong enough to support the loads without deflecting too much. For example, a road bridge will have to carry cars and lorries driving across it, as well as resisting wind forces and movement of its foundations. Some loads on a structure are called static. Other loads are dynamic. Static loads exert a constant force on the structure whereas dynamic loads exert a varying force. In the road bridge example, static loading will result from the weight of the bridge itself. The effect of vehicles crossing the bridge and the wind forces are examples of dynamic loads.

△ A suspension bridge is a frame structure.

There are four types of members commonly used for framed structures:

- beams
- ties
- struts
- columns.

BEAMS

A **beam** is designed to withstand bending. Bending is mainly caused by vertical forces. Beams are often used to span gaps; for example, the roadway part of a suspension bridge.

Beams can be classified by the way they are supported. Three common types are:

- a cantilever beam, which is supported at one end only

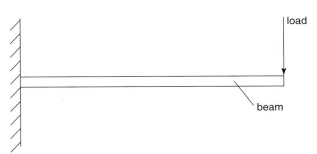

- a simply supported beam, which is supported at both ends

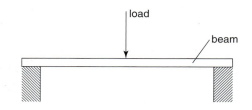

- an overhanging beam, where the load overhangs the end of a beam.

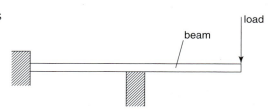

When a beam is bent, one of its surfaces is in tension and the other surface is in compression.

This is an important factor when selecting materials for beam applications. For example, some materials such as concrete and some types of wood are stronger in compression than in tension. A beam designer will need to take these factors into account when considering how best to use the beam in practice.

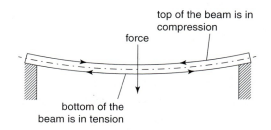

TIES

A **tie** is a member designed to resist tension forces (tensile forces). Ties include rods, bars, guy ropes and suspension bridge cables. Many ties are solid bars. However, they can also be made from cables, wires or rope which 'strengthen up' when stretched.

STRUTS

Struts are members which resist compressive (squeezing) forces. Members made from cables, wires or ropes cannot be used as struts because they are not able to support the load acting on them. Some materials, such as concrete or cast iron, perform better in compression than in tension and are often used to make struts.

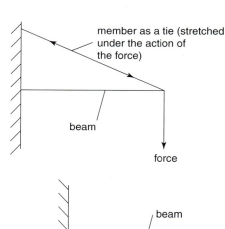

SKILLS ACTIVITY

Using drinking straws or similar materials, make a model of each of the frame structures shown in the diagram. The parts can be joined together with glue, masking tape or other quick assembly methods. Support the frames on a desk or table and press the frames at the loading points shown. When loading, decide whether the members are a tie, strut or beam.

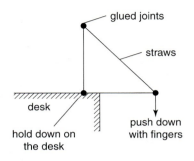
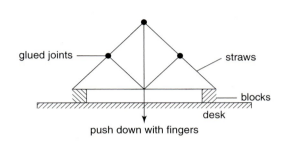

COLUMNS

Columns are perpendicular members used to support beams. Columns resist compressive forces due to the forces pressing down on them. Columns may also have to resist twisting and bending forces. For example, lamp posts need to be able to withstand wind forces acting on them in use.

 REFLECTIVE LOG Make a list of some framework structures in the area in which you live. For each structure, note down the purpose of each part of the framework and the materials it is made from.

 KEY TERMS

beam: a structural member which spans gaps and supports the loads acting on it

column: a pillar designed to support the weight of a structure pressing down on it

structural member: a single component of a structure

strut: a structural member designed to resist compressive forces

tie: a structural member designed to resist tension forces

4·2·3 Strengthening frame structures

LEARNING OBJECTIVES
By the end of this unit you should:
- ✓ understand how triangulation can strengthen a frame structure
- ✓ understand how bracing and gusset plates strengthen structures at their corners.

Frame structures are designed so that each member has a particular purpose. For example, some members will be ties designed to resist the tension forces acting on the structure; others are struts designed to resist compressive forces.

TRIANGULATION
Rectangular shapes can 'collapse' due to the forces acting on them. Connecting a cross member between two corners of the frame will strengthen it. This principle is called **triangulation**, and it is a common principle used to strengthen frame structures. Members used for triangulation can be in tension or compression depending on how the forces act on the frame.

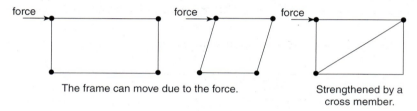

△ The ArcelorMittal Orbit in London's Queen Elizabeth Olympic Park uses a frame structure composed of many triangular shapes to increase its strength.

Cross-bracing and gusset plates are common ways of strengthening up the corners of the frames.

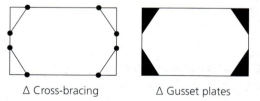

△ Cross-bracing   △ Gusset plates

SKILLS ACTIVITY

Using drinking straws or similar materials, make a rectangular frame structure. Using the principle of triangulation, join the corners with one and then two cross braces. These will connect the corners of the rectangular frame. Again, note what happens when you squeeze the frame with the cross braces in place.

Make a second rectangular frame out of the straws or your chosen material. Strengthen the corners by (i) cross-bracing the corners and (ii) using cardboard to make gusset plates. Note down how these methods strengthen up the rectangular frame.

TRUSSES

Trusses are commonly used for bridges, cranes and the roofs of buildings. For example, roof trusses have beam, tie and strut members joined together to support the roof of a building. Bridge trusses tend to be more complicated. Each member of the truss will either be in tension or compression when loaded and will act either as a tie or strut.

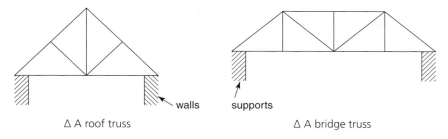

△ A roof truss   △ A bridge truss

SKILLS ACTIVITY

Look around the area where you live and identify three examples of structures where trusses are used. Make a sketch of each truss and label which part is a beam, and which parts might be ties or struts.

SPACE FRAMES

A **space frame** is a three-dimensional truss used to span large areas. Buildings with space frames are built with no intermediate columns to create a large open space. Space frames are typically made from tubular members which help make the frame lightweight.

△ A space frame

KNOWLEDGE CHECK

1. Describe two ways to strengthen a rectangular frame when used as part of a structure. (4)
2. Give two examples of where trusses may be used as part of a building or other construction application. (2)

KEY TERMS

space frame: a three-dimensional lightweight truss structure used to span large areas

triangulation: the use of triangular shapes to give stability to a structure

truss: a structural framework used to support loads, for example in a roof or bridge

4·2·4 Nature of structural members

LEARNING OBJECTIVES
By the end of this unit you should:
- ✓ understand that deflection of a structural member depends on the material it is made from
- ✓ understand that deflection of a structural member depends on its cross-sectional shape
- ✓ understand that struts can fail by buckling
- ✓ understand that deflection of a beam is related to the depth of the beam.

Two factors that affect the performance of ties and struts are (i) the material they are made from and (ii) their **cross-sectional shape**.

For example, concrete is stronger in compression than in tension. When being 'squeezed', concrete can easily withstand the forces exerted on it. However, when used in tension it has a tendency to crumble if the load is too high. If concrete is used in tension it generally requires reinforcing with steel bars. This is called reinforced concrete. When reinforced concrete is used in tension; for example, as a tie member, the reinforcing bars withstand the tension forces acting on it.

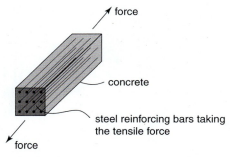

△ Reinforced concrete

A material may also be selected for the following reasons:
- it is lightweight, for example aluminium
- it can be joined easily, for example mild steel can be easily welded
- it may require little maintenance such as painting
- it may have good corrosion resistance.

SKILLS ACTIVITY

Choose three frame structures used for buildings. Identify the materials used to make the frames. Note down why you think each material has been chosen for the structure.

BUCKLING OF STRUTS

Struts fail by **buckling** if the compressive forces are too high. Struts generally need to have larger cross-sectional areas than ties to prevent buckling under a load.

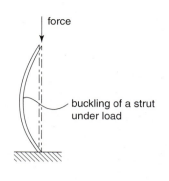

buckling of a strut under load

BENDING OF BEAMS

The deflection and strength of a beam depends on the material it is made from, the distance between supports and its cross-sectional shape. The cross-sectional shape is important in determining how much a beam will bend. For example, a 'thin' beam will deflect much more than a 'deep' beam under the same load.

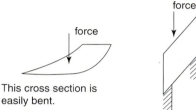

This cross section is easily bent.

A deeper beam is more difficult to bend.

When a beam is bent, its outer surfaces withstand most of the bending forces, so beams can be designed so that they have the least material in their middle section and more material on their outside surfaces. This is the principle of the **I-beam**, which is used for many structural applications. Reducing the material in the middle of the beam reduces the weight of the beam, which is useful for large span applications.

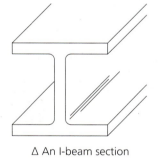

Most of the bending forces are taken on the outer surfaces.

The middle section takes far less of the bending forces.

△ A beam during bending

△ An I-beam section

Composite beams work on the same principle. A composite is made by having a light core material sandwiched between two stronger, skin materials, for example sheet steel. Because the skins are on the outer surfaces of the beam, they withstand most of the bending forces. The matrix material is light and reduces the overall weight of the beam. Composite beams find wide use in transport applications such as aircraft floor panelling.

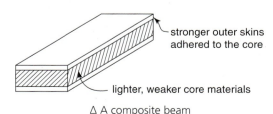

stronger outer skins adhered to the core

lighter, weaker core materials

△ A composite beam

Levers can be used as part of a machine or structure to make a job easier to carry out. An example of this is a paper punch, which is made from a number of shell structures joined together. When the punch is operated a second class lever is used to reduce the effort required to punch through a stack of paper.

SKILLS ACTIVITY

△ A paper punch

1. Identify two examples where an I-beam might be used to span a gap. State why the I-beam has been used.

2. Identify two examples of where a composite beam might be used for transport applications. Again, give reasons why the composite beam has been chosen.

TYPICAL MEMBER CROSS-SECTIONS
Below are examples of typical member cross-sections.

solid square   solid rectangle   rectangular tube   solid round   round tube   angle   i-section

KNOWLEDGE CHECK
1. What type of member might fail by buckling? (1)
2. What happens at the surfaces of a beam when it is bent? (2)
3. Give one use of an I-beam and one use of a composite beam, and explain why they are used for these applications. (4)

KEY TERMS

buckling: a type of failure in which a component bends out of shape, usually as the result of pressure or heat

composite sandwich beam: a beam constructed by a lightweight material sandwiched between two stronger materials

cross-sectional shape: a shape made by cutting through something, for instance ties, struts, columns and beams

I-beam: a girder beam which has an I-shaped cross-section with two flanges at the top and bottom of the beam

4·2·5 Applied loads and reactions

LEARNING OBJECTIVES

By the end of this unit you should:
✓ understand that point loads on a beam can be shown by an arrow
✓ understand that the upward reaction forces are equal to the downward forces acting on a beam
✓ be able to determine the reaction forces for a simple beam system.

REACTION FORCES AND LOADS

A reaction is the upward force applied to a beam when it rests on its supports. **Reaction forces** tend to be exerted by foundations, columns or walls holding up the beam. To help calculate reaction forces, a beam is assumed to be in static equilibrium. This means that the downward forces acting on the beam are equal to the reaction forces acting on the beam. The reaction force acts in the opposite direction to the load.

Reaction forces and loads are assumed to be **point loads.** These are loads which act at a single point on the beam and are shown by arrows.

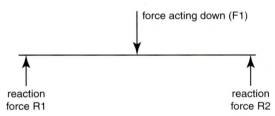

F1 = downward force acting on the beam

R1 = left side reaction force

R2 = right side reaction force

F1 = R1 + R2

DOWNWARD FORCES EQUAL UPWARD FORCES
Example 1

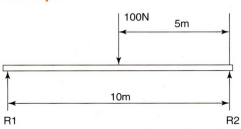

A beam has a point load of 100 N acting at its centre. Since the load is at the centre, each reaction (R1 and R2) will support half of this load. The load is therefore shared between the two reactions R1 and R2, so reaction R1 = 50 N and reaction R2 = 50 N.

Example 2

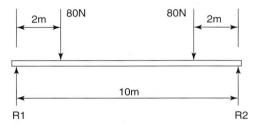

A beam has two point loads which are both 80 N. They are positioned at the same distance (2 m) from reaction points R1 and R2. The two downward forces will, again, be shared between the two reaction forces because they are positioned symmetrically along the beam. Therefore:

R1 = (80 N + 80 N) ÷ 2 = 80 N

R2 = (80 N + 80 N) ÷ 2 = 80 N

SKILLS ACTIVITY

What will be the values of reaction force R1 and reaction force R2 in the example shown?

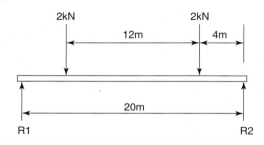

UNIFORMLY DISTRIBUTED BEAMS

A beam can also have a load which is distributed evenly along its length, such as where materials are stacked on floors or shelves.

Uniform loads are given as a force per metre, for example 200 N/m. This is assumed to act all the way along the beam. A distributed load is often shown as a series of arrows on top of the beam.

△ A shelf is a common example of a uniformly distributed beam.

Example

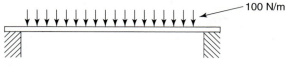

△ A distributed load

In this diagram, the beam has a distributed load of 100 N/m. If the beam has to support a total load over three metres:

Total load = 3 × 100 N = 300 N.

The reaction values for a uniform beam of this type can be calculated by assuming that the total load, for example 300 N, acts at the centre of the beam and as a single point case. Therefore, each reaction (R1 and R2) is 300 N ÷ 2 = 150 N.

SKILLS ACTIVITY

1. A beam that is 10m long has a uniformly distributed load of 500 N/m acting on it. What is the total load acting on the beam?

2. Determine the reaction forces (R1 and R2) for the uniformly distributed load beam shown in the diagram.

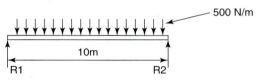

KNOWLEDGE CHECK

1. What is meant by the term *reaction*? (1)
2. A simply supported beam has a load of 400 N acting on it. If one of the reactions is 100 N, what will be the value of the other reaction? (2)
3. A simply supported beam has a length of 8 metres. If the beam supports a force of 2kN acting at its centre, what will the value of the two reaction forces be? (2)

KEY TERMS

applied load: the size of a load acting vertically on a beam

Newton: the unit of force which has the symbol N

point load: a load which is applied to a particular point on a beam

reaction force: the force exerted by the beam supports which act to resist the applied load

uniformly distributed load: a load which is distributed equally along the length of the beam

4·2·6 Moments

LEARNING OBJECTIVES
By the end of this unit you should:
✓ understand that a moment is a turning force
✓ understand the equation M = F × d
✓ be able to use the principle of moments to determine the reactions of non-symmetrical beams.

WHAT IS A MOMENT?

A **moment** is a turning force. The forces exerted by levers, spanners and door handles are all examples of **moments of forces**.

The size of a moment acting on an object, such as a lever, depends on the magnitude (size) of the force and its distance from its pivot point.

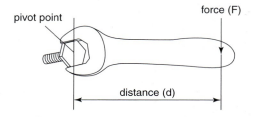

The size of a moment =
Force (in Newtons) × distance (in metres)

M = F × d

The units of moments are Nm.

Example
An 80 N force acts 2 metres from the pivot point.

Moment of force = Force × distance from the pivot point
$$= F \times d$$
$$= 80 \times 2$$
$$M = 160 \text{ Nm}$$

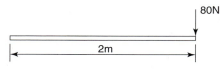

SKILLS ACTIVITY

The force acting on a lever and the distance from its pivot point are given in the table. Determine the moment of force for each case.

Force acting on the lever	Distance (m) from the pivot point	Moment of force (Nm)
450 N	0·25	
1.2 kN	2·1	
810 N	0·75	

USING THE PRINCIPLE OF MOMENTS FOR NON-SYMMETRICAL LOADS ON BEAMS

There are many situations where the loads acting on the surface of a beam are not symmetrical along the beam's length. For example, in the beam shown opposite the load is not acting at the centre of the beam (symmetrically positioned). Instead, it is situated nearer to the left-hand reaction (R1). As such, reaction R1 will have to support proportionally more of the 2000 N load than is supported by the R2 reaction. The principle of turning moments can be used to determine the values of R1 and R2 in non-symmetrical beam loading cases such as these.

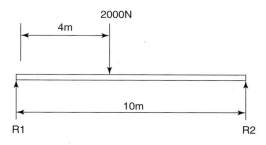

To apply the principle of moments, it is first assumed that the beam is free to turn around one of the reaction points due to the influence of the load and other reaction forces acting on the beam. For example, if a beam is free to turn around reaction point R1 then there will be two moments of force acting on the beam: (i) a clockwise moment which is equal to the 2000 N load × 4 m, and (ii) an anticlockwise moment which is equal to the unknown moment R2 × 10 m. For the beam to be in static equilibrium under the loading condition, the clockwise moment acting on the beam must equal the anticlockwise moment.

Example 1

In the example below, using the diagram above, the values of R1 and R2 can be calculated as follows:

Anticlockwise moments acting on the beam = clockwise moments acting on the beam

$$R2 \times 10\ m = 2000\ N \times 4\ m$$
$$R2 \times 10\ m = 8000\ Nm$$
$$R2 = \frac{8000\ Nm}{10\ m} = 800\ N$$

To calculate the magnitude of R1, you can use the principle that upward forces acting on the beam must equal the downward forces acting on the beam. Therefore:

$$R1 + R2 = 2000\ N$$
$$R1 = 2000\ N - R2$$
$$R1 = 2000\ N - 800\ N$$
$$R1 = 1200\ N$$

As you can see, the left-hand reaction (R1) has to withstand much more of the load because of the load's position towards the left-hand side of the beam.

Examples which have more than one non-symmetrical load can be analysed in a similar way.

Example 2

The principle of moments can be used in a similar way to determine the magnitude of R1 and R2 in an example where the beam has two non-symmetrical point loads.

Assuming that the beam pivots around the R1 reaction:

Anticlockwise moments acting on the beam = R2 × 5 m

Clockwise moments acting on the beam = (2000 N × 1m) + (1000 N × 3 m)

$$\text{Anticlockwise moments} = \text{clockwise moments}$$
$$R2 \times 5\,m = (2000\,N \times 1\,m) + (1000\,N \times 3\,m)$$
$$R2 \times 5\,m = 2000\,Nm + 3000\,Nm$$
$$R2 = \frac{5000\,Nm}{5\,m}$$
$$R2 = 1000\,N$$

For R1:
$$R1 + R2 = 3000\,N$$
$$R1 = 3000\,N - 1000\,N$$
$$R1 = 2000\,N$$

SKILLS ACTIVITY

For the two non-symmetrical examples shown below, calculate the value of the reactions R1 and R2.

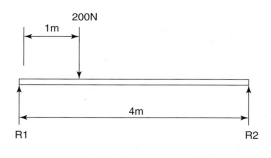
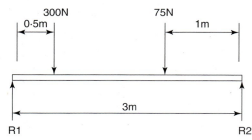

> **KEY TERMS**
>
> **moment:** a turning force around a pivot
>
> **moment of force:** the magnitude of a force multiplied by its distance from the pivot point, expressed as the unit Nm

4·2·7 Materials

> **LEARNING OBJECTIVES**
>
> By the end of this section you should:
> - ✓ understand that materials properties can be divided into mechanical and physical properties
> - ✓ be able to define the main mechanical properties such as tensile strength, compressive strength and toughness.

It is important to select suitable materials for structural parts. For example, some materials are good in tension and some are good in compression. Others have good corrosion resistance properties and some are lightweight. In many cases, a material might be chosen because it has good aesthetic properties.

Materials are selected based on their specific properties, and the properties of materials can be categorised into mechanical or physical properties.

In order to improve their working characteristics, materials can be combined: for example, concrete can be reinforced with steel rods or plastic flakes to improve its resistance to tension. Softwood can be reinforced by laminating, which is also a method of increasing length and depth of a wooden beam. Laminated beams are used in many buildings for their aesthetic properties.

Stone is often bonded into a pattern in the same way that bricks are. This helps to prevent cracks running either vertically or horizontally along the weaker mortar joint.

Metals are used both for their structural qualities and for their aesthetic qualities. As a roofing material, copper is very resistant to corrosion, but will also develop a green colouring, known as verdigris. Stainless steel has the ability to resist most forms of chemical corrosion present in the atmosphere. It is particularly well-suited to specific locations, such as swimming pools or food production areas.

MECHANICAL PROPERTIES

Make sure to consider mechanical and physical properties carefully when choosing structural parts. The following tables list some you may consider.

Tensile strength	the ability to withstand pulling (tension forces)
Compression strength	the ability to withstand squeezing forces
Shear strength	the ability to withstand forces which 'shear' a material apart
Torsional strength	the ability to withstand a twisting force
Bending strength	the ability to withstand bending forces
Hardness	the ability to withstand scratching and wear
Elasticity	a measure of how flexible a material is
Toughness	the ability to absorb energy and deform without fracturing

PHYSICAL PROPERTIES

Resistance to corrosion	rusting is the main form of corrosion in structural steel members
Thermal conductivity	how well a material conducts heat
Optical properties	whether a material is transparent, translucent or opaque

Material class	Example	Tensile strength	Compressive strength	Shear strength	Torsional Strength	Bending strength	Hardness	Elasticity	Toughness	Resistance to corrosion / rot	Thermal conductivity	Electrical conductivity	Optical properties
Hardwood	Teak	5	4	4	4	4	4	2	5	5	1	✗	Opaque
Softwood	Pine	4	3	4	4	3	2	2	2	1	2	✗	Opaque
Ferrous metal	Steel	5	4	4	4	3	4	3	4	2	4	✓	Reflects
Non-ferrous metal	Copper	3	3	3	2	2	3	2	3	4	5	✓	Reflects
Alloys	Brass	3	3	4	3	4	3	3	4	4	4	✓	Reflects
Stone	Granite	2	5	4	1	3	5	1	5	4	1	✗	Reflects
Concrete	Not reinforced	1	5	3	1	1	5	1	5	5	2	✗	Opaque
Concrete	Reinforced	3	5	4	2	4	5	3	5	5	2	✗	Opaque
Thermoplastics	Polypropylene	2	3	3	4	5	2	3	4	5	1	✗	Opaque
Thermosets	Urea formaldehyde	3	4	3	2	2	3	2	3	5	2	✗	Opaque
Composites	Carbon Fibre	4	4	4	4	4	4	3	5	5	2	✗	Opaque

1 = low
5 = high

4·2·8 Testing   EXTENDED

LEARNING OBJECTIVES
By the end of this section you should:
- ✓ understand how dial gauges and strain gauges are used for testing structural parts
- ✓ understand that strain gauges form part of a measuring system
- ✓ be able to read a dial gauge to measure structural deflections.

STRUCTURAL TESTING

It is important to be able to test the strength and deflection of structural parts. For example, a roof beam may bend too much when loaded, or a crane cable might extend beyond its intended length. Two devices commonly used to test the deflection and extension characteristics of structural parts are dial gauges and strain gauges.

DIAL GAUGES

Dial gauges are convenient ways of measuring small deflections in beams or other structural members. Dial gauges have a dial display (commonly called a clock) where a needle indicates the deflection which has occurred. The gauge changes linear movement to the rotary movement of an arrow on a dial. It can measure very small amounts and is normally calibrated in hundredths of a millimetre (0.01 mm) or thousandths of an inch (0.001 inch). The dial gauge is positioned so that it touches the item being tested and it is observed while loading takes place. Common places to test would be at the end of a cantilever beam or in the centre of a beam supported at both ends.

△ A dial gauge

STRAIN GAUGES

Strain gauges are small electronic devices which are used widely in industry and research for measurement applications, such as indicating the extension of structural members. They are also used in equipment such as load cells, torque meters and extensometers.

The strain gauge is a foil pattern with two electrical connections on an insulated backing. This backing is attached to the test piece using either epoxy resin or cyanoacrylate (superglue). When the structure being tested is put under load, the strain is measured using the slight change in resistance in the foil pattern. Strain gauges are often used in a 'Wheatstone bridge' electronic circuit which uses the change in resistance to vary a voltage that can then be measured. Strain gauges are often used in buildings or on bridges to measure the strain caused by a load.

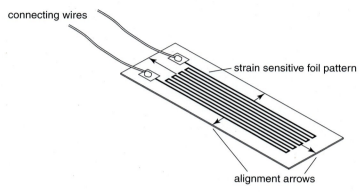

△ A strain gauge

SKILLS ACTIVITY

Think about how you might use strain gauges to measure the extension of a column supporting a building in your school. How many strain gauges might you use on the column? Where might you position them? What other equipment would you need to measure the output of the gauge or gauges?

4·2·9 Joints in structures

LEARNING OBJECTIVES
By the end of this section you should:
- ✓ understand that joints can be rigid or moveable
- ✓ be able to select suitable jointing methods for wood, steel and concrete
- ✓ be able to recognise how cross-bracing and the use of lugs and ribs make joints stronger.

Joints for structures can be used to:
- join members together
- allow a structure or parts of structure to be portable
- carry loads without breaking.

Most joints depend on the type of materials being used and for most materials they can be either temporary or permanent. For example:

Material	Joint
Wood	screws, adhesive, coach bolts
Steel	welding, brazing, riveting, nuts and bolts
Plastics	welding, nuts and bolts, screws
Pre-cast concrete	mechanical connections
Cast concrete	all-in-one monolithic connections

Joints can be moveable, such door or cabinet hinges, which allow parts of the joint to rotate relative to each other.

Joints can also be rigid, for example, the joints between the structural members of a bridge or a welded joint on a car body.

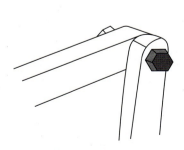

△ A rigid joint

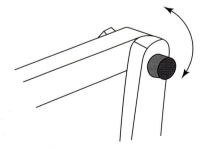

△ A moveable joint

TYPES OF JOINTS

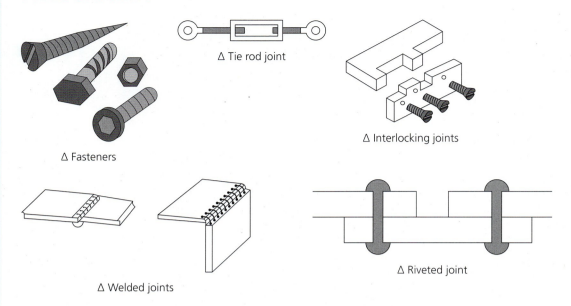

△ Fasteners
△ Tie rod joint
△ Interlocking joints
△ Welded joints
△ Riveted joint

JOINING SOLID AND HOLLOW CROSS-SECTIONAL MATERIALS
Metal joints can be made by brazing, soldering, welding or using machine screws and rivets.

Brazing
Brazing is a common method of joining two materials by melting a filler metal between the two parts being joined. Brazing forms very strong, permanent joints. The filler metal used for brazing needs to have a lower melting point than the materials being joined. (See section 3·10 for more information on brazing.)

Soldering
There are two methods of soldering. These are (i) soft soldering, which is used to make permanent joints between metals such as brass, tin-plate or copper. It is commonly used for electronic circuits, and (ii) hard soldering, which is used for stronger joints. (See section 3·10 for more information on soldering.)

Welding
This is where metal parts are melted along their joints and fuse together as they cool. A filler rod is often used as part of the welding process. (See section 3·10 for more information on welding.)

Machine screws

Machine screws are a special type of screw used for joining metal parts together. The screws fit into holes which have been drilled and tapped for the size of the machine screw. (See section 3·10 for more information on machine screws.)

Riveting

Riveting is used to join metal and other materials, such as plastics. The rivet is placed in a hole drilled through both pieces of material, and its end shaped by hammering or other shaping methods. Rivets are often dome shaped or countersunk when they are finished. Pop rivets are a type of rivet commonly used to join thin sheets of material together. (See section 3·10 for more information on riveting.)

SKILLS ACTIVITY

From the environment around you, identify structures which are held together by (i) a fastener, (ii) a tie rod and (iii) a welded joint. Make a sketch of each joint and briefly describe why you think the joint has been chosen for its purpose.

4·2·10 Forces

LEARNING OBJECTIVES

By the end of this section you should:
- ✓ know the meaning of the terms stress, strain, bending, torsion and shear
- ✓ be able to calculate simple stress values
- ✓ be able to calculate simple strain values
- ✓ be able to recognise shear and torsional forces acting on structural parts and members
- ✓ be able to identify the key features of a stress-strain graph
- ✓ understand the use of safety factors in the design of structural members
- ✓ be able to use the triangulation of force method to determine unknown forces in a simple framework
- ✓ be able to use the parallelogram of force method to determine the resultant of two forces acting on a part.

FORCES

All structures or structural parts need to be able to withstand the forces acting on them. For example, a floor in a house should not deflect too much or completely collapse when a person walks across it.

The size of forces range from relatively small values such as those used to hold a pencil when drawing, to the large forces acting on an airplane when flying at altitude. The SI unit of force is the Newton (abbreviated N). In many cases it is more convenient to use larger units. These include kilonewtons (kN) and meganewtons (MN). A kilonewton is equal to 1000 N and a meganewton is equal to 1,000,000 N.

Two ways that a force can act on a structure or structural part are in tension or in compression. A tension (or tensile force) has the effect of pulling a material apart. A compressive force has the effect of squeezing a material together. It is important to be able to determine the effect of forces on structural parts. This helps predict whether the structure will be strong enough for the job it is intended for. The concepts of **stress** and **strain** help predict whether a structural component will be able to withstand the effect of such forces.

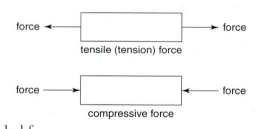

STRAIN

A structural part, such as a tow bar when towing a car, will extend slightly due to the tensile force pulling it apart. The amount of extension will depend on the size of force and also the material the part is made from. A structural part can also shorten due to a compressive (squeezing) force acting on it. For example, a chair leg will shorten a certain amount when someone sits on the chair.

△ A car tow bar

Some extensions or compressions will be relatively small due to the applied force. For example, a steel column holding up the roof of a school building will shorten slightly because it is made from steel. Others can be relatively large, such as when an elastic band is being stretched by hand. This change in length compared to a material's original length is called strain; this is determined from the relationship:

$$\text{strain} = \frac{\text{change in length}}{\text{original length}}$$

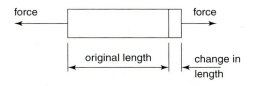

The strain can be tensile if the force pulls the material apart, or compressive if it squeezes it together. It is important to realise that strain is the ratio of the two lengths involved and as such does not have any units.

Example

A solid bar used for towing a car has an original length of 1.5 metres. If the bar extends by 2 millimetres when used for towing then the magnitude of strain can be found by

$$\text{strain} = \frac{2}{1500} = 0.0013$$ [note that the values have both been changed to metres and that there are no units associated with strain].

SKILLS ACTIVITY

There are many examples where structural parts and equipment might extend or shorten a certain amount during use. A few examples are given in the table below. For each example, use the intensity of strain formula to calculate the strain using the original length and change in length. Indicate whether you would expect the strain to be tensile or compressive.

Product	Original length (m)	Extension (mm)	Value of strain
A climbing rope	60	30	
A chair leg	1·2	0·5	
A lift cable	30	4	

STRESS

In addition to extending or compressing a material, a force will have an effect on the overall strength characteristics of a material. For example, if an applied force is too high, a component may break due to the size of the force being applied. The concept of stress is used to help calculate whether a structural part will fail either by (i) completely breaking under an applied force or (ii) deemed to have failed because it has permanently stretched, compressed or bent during use, making it unfit for purpose as a result. For example, it would be a poor design if a bicycle frame became permanently twisted as soon as a cyclist sat on the saddle.

The value of stress depends on the size of the force being applied and also the cross-sectional area over which the force acts. As with strain, stress can be tensile or compressive. The value of stress can be determined from the relationship:

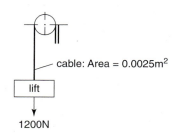

$$\text{stress} = \frac{\text{force}}{\text{cross sectional area}} \quad [\text{units N/m}^2]$$

Example

A lift cable with a cross-sectional area of 0.0025 m² is acted on by a tensile force of 1200 N. What is the stress in the cable?

$$\text{stress} = \frac{\text{force}}{\text{cross sectional area}}$$

$$\text{stress} = \frac{1200}{0.0025}$$

$$\text{stress} = 480000 \text{ N/m}^2 \text{ or } 480 \text{ kN/m}^2$$

SKILLS ACTIVITY

Determine the stress for the following examples:

1. A tie bar in a bridge structure has a cross-sectional area of 12 mm² and resists a pulling force 2000 N.
2. A link pin has a cross-sectional area of 0.030 m². If a compressive force of 800 N squeezes the link pin, what is the value of stress exerted in the pin?

SHEAR AND TORSIONAL FORCES

Many structural and mechanical parts have to be able to withstand shear and torsional (twisting) forces. As such, designers need to consider these factors when selecting parts for a particular purpose. Torsion describes the twisting effect of a structural or mechanical part due to a torque being applied to it. Parts such as axles, motor drive shafts, screwdriver blades and bolts have to have good torsional properties otherwise they might fail when being used. In most cases, when parts completely break due to torsion, they will break because the two surfaces shear in a twisting fashion.

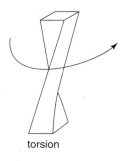

torsion

The shear strength of a material part is related to how well it can resist the shear forces acting on it. Shear forces tend to produce a sliding type of failure, as opposed to the part breaking in a purely tensile (pulling) way, where the fracture surface is at right angles to the force applied to it. Designers need to consider shear properties when selecting materials for mechanical and structural parts, so that they do not fail when used. Some materials, such as concrete, are weak in tension. If the load is too high for the concrete to withstand, it will tend to fail by shearing at 45°.

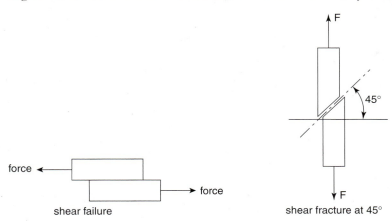

STRESS-STRAIN GRAPHS EXTENDED

A useful way to obtain data about the way structural components behave under the action of tensile forces is to use a **stress-strain graph**. Stress-strain graphs are obtained from the results of a tensile test. During the test a specimen with a known cross-sectional area and original length is pulled apart on a tensile testing machine until the material fails.

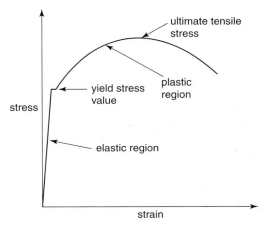

△ A stress-strain graph

A typical stress-strain graph is divided into two main parts. These are called (i) the elastic region and (ii) the plastic region. There are two other important points on the graph called the yield stress and the ultimate tensile stress.

If a structural component is pulled beyond its yield stress, the material is said to have yielded and a certain amount of permanent extension will take place. This permanent extension is also called plastic deformation. However, if the material is stressed below its yield stress and the force removed, it will return to its original size. When this happens there is no permanent extension and the material is said to have behaved elastically. In practice many structural parts and components, such as suspension bridge cables and turbine blades in aircraft engines, are designed to be used below their yield stress so that there is no permanent change in dimensions when being used.

The point at the top of the stress-strain graph is called the ultimate tensile stress of the material. This is the maximum stress the material can withstand before it finally breaks apart.

FACTOR OF SAFETY

When designing a structural part it is essential that the stress never exceeds the ultimate tensile stress or the yield stress if it is not to have some permanent deformation. It is common practice to use a **factor of safety** to ensure that the structure or structural component stays within the safe design limits it is intended for.

A factor of safety takes into account many of the uncertainties that may occur during the product's use, such as changes in material characteristics, structural overloading and flaws that may have occurred during manufacture. For example, a lift cable is often designed to withstand a force which is ten times greater than the nominal design force. This means that it has a safety factor of × 10.

DETERMINING SIMPLE FORCES ACTING ON A STRUCTURE

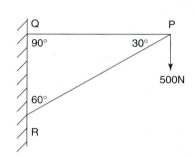

In many cases, it is necessary to determine the magnitude of forces acting on particular structural members. For example, in the simple three-member frame shown opposite it might be necessary to determine the forces acting on members PQ and PR which form the basis of the frame.

USING TRIANGULATION OF FORCES OR PARALLELOGRAM METHODS TO DETERMINE FORCES

EXTENDED

The forces which act on the various members of structures are vector quantities and as such their magnitudes can be solved by graphical means. Two basic graphical methods are: (i) using a triangulation of forces method to determine the relationship between three forces in static equilibrium and (ii) using a parallelogram method for two forces in static equilibrium. Both methods use scaled diagrams to help determine any unknown forces in the structural framework.

Triangulation of forces

An example of how to make use of the method is to consider the simple three member structure shown opposite. As you can see, the frame is supporting a 500 N force and there is a need to determine the magnitudes of force in the tie and strut members. Because there are three forces involved, you can use a triangulation of forces method.

A suitable scale is selected for the forces involved and then each force vector is drawn to scale to form a closed triangle, which represents the balanced forces.

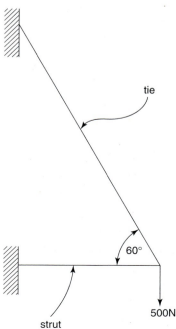

The method is as follows:

(i) Select a suitable scale for the known force, for example, 1 cm = 50 N

(ii) Draw the vector to scale: 10 cm in a downward direction

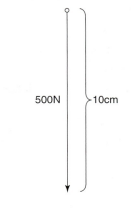

(iii) Using a protractor and rule, draw construction lines to 'map in' the tie and the strut vectors. Note that they need to be drawn at the same angles as represented on the original framework diagram. For this example, the tie is at an angle of 60° to the horizontal and the strut is horizontal. Each vector will follow on from the other in a 'top to tail' fashion.

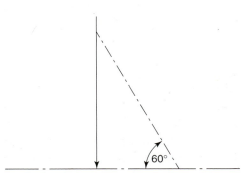

(iv) Inspect the diagram and complete the triangulation of forces diagram for the members.

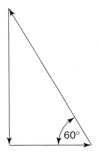

(v) Using a rule, measure the lengths of the tie and strut vector representations and use the scale to determine the magnitude of each of the forces.

Parallelogram of force method

This method is useful for determining the **resultant** (equivalent single force) of two forces acting on a body.

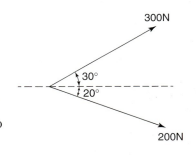

The method, again, makes use of the principle that vectors can be added together in a 'top to tail' fashion to help determine the resultant for the two forces. Here is an example.

The method is quite simple:

(i) Select a suitable scale for the forces acting on the members. Because the known forces are 300 N and 200 N, a scale of 2 cm:100 N would be suitable.

(ii) Using a rule and protractor, draw one of the forces to scale and at the right angle on paper.

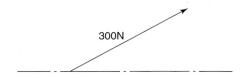

(iii) Draw the second vector to scale and at the correct angle from the 'end' of the first vector. This uses the top to tail method of adding vectors together.

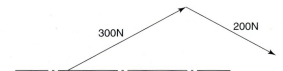

(iv) Complete the parallelogram of forces. The resultant force is the force which 'stretches across' the corners of the parallelogram.

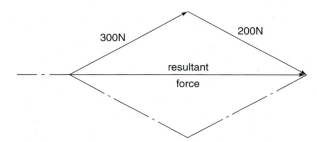

(v) Measure the length of the resultant and use the scale to determine the magnitude of the resultant force.

> **KEY TERMS**
>
> **factor of safety:** the ratio of the ultimate breaking strength of a material against actual force exerted on it
>
> **resultant:** a single force that has the same effect as all the other forces acting on an object
>
> **strain:** the amount of extension or compression of a material divided by its original length
>
> **stress:** the force acting on a material divided by its cross-sectional area
>
> **stress-strain graph:** a graph showing the relationship between stress and strain for a material under load

4·3 Mechanisms
4·3·1 Basic Concepts

LEARNING OBJECTIVES

By the end of this unit you should:
- ✓ understand that mechanisms are made up of parts like gears, levers, springs, cranks, and pulley and belt drives
- ✓ understand mechanical advantage
- ✓ understand that mechanisms are not 100 per cent efficient.

WHAT ARE MECHANISMS?

Mechanisms are made up of moving parts such as gears, **levers**, springs, cranks, and pulleys and belt drives. Many mechanisms are designed to convert one type of motion to another, such as rotary motion to linear motion. For example, a bicycle transfers the rotary motion of the peddling action to the linear motion of the bicycle as it moves forward. Mechanisms are used to make a job easier to do.

△ Bicycles turn rotary motion into linear motion.

Basic machine elements, for example levers, pulleys and gears, can be regarded as the 'building blocks' for all mechanisms and machines. However, no mechanism or machine is 100 per cent efficient and performance can be lost through friction, and slippage and sliding during operation. Energy can also be lost through heat and sound. There are a number of terms and concepts used to determine the effectiveness of each 'building block'. These include **load**, **effort**, **fulcrum**, **mechanical advantage**, **velocity ratio** and **efficiency**.

MECHANICAL ADVANTAGE, VELOCITY RATIO AND EFFICIENCY

Mechanical advantage, velocity ratio and efficiency are concepts which describe the characteristics of mechanisms in input and output terms.

Mechanical advantage relates to the ease with which mechanisms can lift loads, transmit power and produce useful work. For example, if a pulley is used to lift a load then the mechanical advantage is the advantage gained by using the pulley in numerical terms (for example, a mechanical advantage of 4 times the effort used).

Velocity ratio, on the other hand, relates to the relative motion and velocities of parts making up the mechanism. Velocity ratio is particularly useful when determining the output speeds of belt or gear drives when the input speed is known. Velocity ratio is also used for determining the relative distances that mechanism parts move when operating.

Efficiency is a measure of the effectiveness of mechanisms in energy transfer terms and in practice, no machine is 100% efficient. Performance can be lost through heat, friction, noise, slippage and other such losses. Mechanical advantage, velocity ratio and efficiency can be described as

$$\text{Mechanical advantage (MA)} = \frac{\text{output force or load}}{\text{input force or load}}$$

$$\text{Velocity ratio (VR)} = \frac{\text{distance moved by effort}}{\text{distance moved by load}}$$

$$\text{Efficiency} = \frac{\text{output work or power}}{\text{input work or power}} \times 100\%$$

LOAD, EFFORT AND FULCRUM

The terms load, effort and fulcrum relate to the operation of levers. Levers are the basis of many mechanisms, such as bicycle brake levers and scissors. Levers can be used to move weights with the least amount of effort. The fulcrum is the point where the lever pivots.

The effort is the point where force is applied to a lever. The load is the point where the force being moved is placed.

The position of the fulcrum relative to the effort and load determines the class of lever.

- If the fulcrum is between the effort and load, it is a Class 1 lever (sometimes called a first order lever).
- If the load is between the effort and the fulcrum, it is a Class 2 lever (or second order lever).
- If the effort is between the fulcrum and load, it is a Class 3 lever (or third order lever).

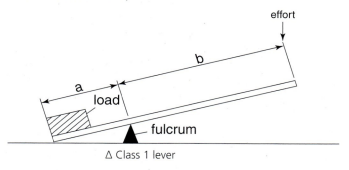

△ Class 1 lever

The ease with which a load can be lifted using a lever is related to its **mechanical advantage**. Class 3 levers do not provide a mechanical advantage, but they can increase the precision of the movement. The amount of mechanical advantage provided by Class 1 and Class 2 levers will depend on the relative distances between effort, fulcrum and load. If there is a greater distance between effort and fulcrum than between load and fulcrum, there will be a higher mechanical advantage for the lever.

The mechanical advantage of a lever is:

$$\text{Mechanical Advantage (MA)} = \frac{\text{length } a}{\text{length } b}$$

Class 1 levers

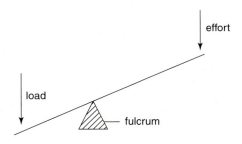
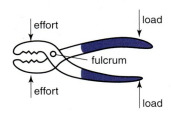

Class 2 levers

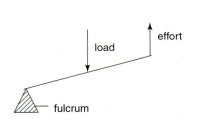
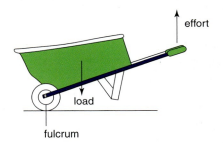

Class 3 levers

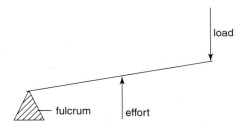
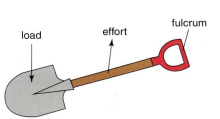

Example

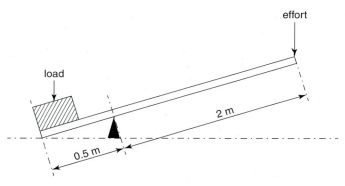

A lever has a length *a* of 2 m and a length *b* of 0.5 m. What is the mechanical advantage of the lever? What will be the effort required to lift a force of 80 N?

$$\text{MA} = \frac{\text{distance } a}{\text{distance } b}$$

$$= \frac{2}{0 \cdot 5}$$

$$= 4$$

The effort required to lift the 80 N load $= \frac{80 \text{ N}}{4}$

$$= 20 \text{ N}$$

SKILLS ACTIVITY

The lever shown in the diagram has a length a of 1.5 m and length b of 0.5 m. Determine the mechanical advantage of the lever. What effort is needed to raise a load of (i) 300 N and (ii) 480 N?

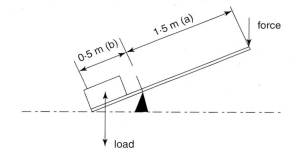

LINKAGES

Where parts of a mechanism need to be joined together, **linkages** can be used. In many cases, linkages are simple bars with joints at either end which allow movement. The joints will often include bearings, which reduce wear on the joint and reduce friction, allowing links to operate quietly without any loss of efficiency.

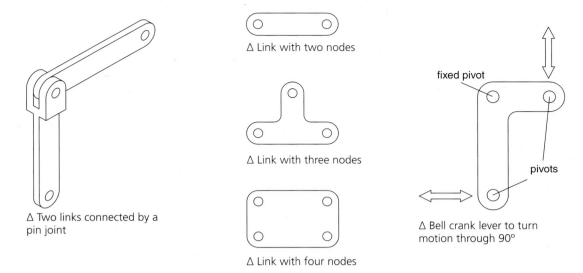

△ Two links connected by a pin joint

△ Link with two nodes

△ Link with three nodes

△ Link with four nodes

△ Bell crank lever to turn motion through 90°

PULLEYS

The mechanical advantage of a pulley system is equal to the number of pulleys in the system.

VELOCITY RATIO OF A BELT AND PULLEY SYSTEM

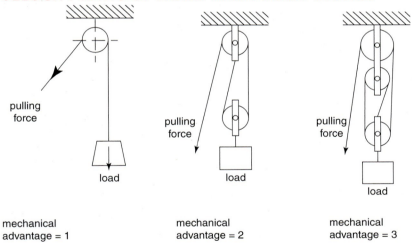

mechanical advantage = 1

mechanical advantage = 2

mechanical advantage = 3

The velocity ratio is useful when determining the output (or input) speeds of gear or pulley systems, if the speed of the driver or driven gear or pulley is known.

This is the ratio of distance travelled by the effort to the distance travelled by the load.

The formula is normally given by: $\dfrac{\text{distance moved by effort}}{\text{distance moved by load}}$

In a pulley system, the formula for the velocity ratio is:

$\dfrac{\text{diameter of the driven pulley}}{\text{diameter of the driver pulley}}$

In the system shown opposite the velocity ratio is
150 ÷ 50 = 3

The output speed can then be calculated by dividing input speed by the velocity ration (VR):

200 ÷ 3 = 66·6 revolutions per minute (rpm)

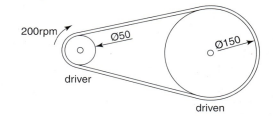

SKILLS ACTIVITY

(i) A pulley system has a driver gear of 200 mm and a driven gear of 800 mm. What is the velocity ratio of the system? If the driver pulley is attached to a motor running at 50 rpm, what will be the speed of the driven pulley?

(ii) The diameters of the driven and driver speeds of three belt drives are given in the table below. For each system determine the speed of the driven gear given each motor speed.

Diameter of driver pulley	Diameter of driven pulley	Motor speed (rpm)
400 mm	1200 mm	500
250 mm	0·5 m	300
40 cm	100 mm	50

VELOCITY RATIO AND GEARS

For two gears meshing, the velocity ratio is related to the number of teeth on each gear and is given by

$$\text{Velocity ratio} = \frac{\text{number of teeth on the driven gear}}{\text{number of teeth on the driver gear}}$$

The velocity ratio of a gear system is often referred to as the gear ratio of the gear system.

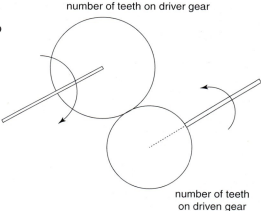

SKILLS ACTIVITY

(i) A driving gear has 90 teeth while the driven gear has 30 teeth. What will be the velocity ratio (gear ratio) of the simple gear system? If the driver gear rotates at a speed of 600 rpm, what will be the speed of the driven gear?

(ii) The output speed of a driven gear when two gears are meshing needs to be 400 rpm. Determine the number of teeth needed on the driven gear if the driver gear has 30 teeth and rotates at a speed of 200 rpm?

EFFICIENCY

This is the ratio of useful work completed by a machine (or output) compared to the effort put into the machine (or input). A perfectly efficient machine will have 100% efficiency and the output will equal the input. This can never be achieved in practice though. Losses through heat, caused by friction, and sound caused by parts rubbing together will reduce the efficiency.

Efficiency can be increased by using bearings on joints and by lubrication of the bearings. In many cases the lubrication will be either oil or grease. This may not be practical though as both of these lubricants will attract dust and dirt which could cause damage. Dry lubrication such as graphite powder can be used to good effect on locks and parts that must remain clean.

Calculating efficiency

Efficiency is a comparison of the useful work provided by a machine to the work supplied to the machine. Mathematically this can be written as:

$$\text{Efficiency} = \frac{\text{work or power out}}{\text{work or power in}} \times 100\%$$

Example

A gear box is 80% efficient. If the input power to the gear box is 1200 watts, what will the output power of the gear box be?

$$\text{Efficiency} = \frac{\text{output power}}{\text{input power}} \times 100\%$$

$$\text{Output power} = \frac{80}{100} \times \text{input power}$$

$$\text{Output power} = 0 \cdot 8 \times 1200 = 960 \text{ W}$$

KNOWLEDGE CHECK

1. Draw a lever and label the fulcrum, load and effort. (1)
2. Explain what is meant by the term *mechanical advantage*. (2)
3. A simple gear train has a driving gear with 100 teeth and a driven gear with 20 teeth. If the output gear needs to have a speed of 300 rpm, what input speed is required? (4)
4. A machine is rated as 75% efficient. If the input power to the machine is 2.5 kW, what will be the output power delivered by the machine in watts? (4)

KEY TERMS

efficiency: the ratio of useful work or power to actual input work or power

effort: the force applied to the input part of a mechanism

fulcrum: the pivot point about which a lever turns

lever: a mechanism used to move loads which will rotate around a fulcrum

linkage: a set of rigid parts joining two or more components of a mechanism together. Each link has at least two nodes or attachment points.

load: the force that is supported or moved by a mechanism

mechanical advantage: the advantage gained by using a tool, mechanism or machine

mechanism: a device which uses moving parts to take an input motion and force, and converts it to a different motion and force

velocity ratio: the ratio of distances moved by the effort and load at the same time

4·3·2 Conversion of motion

LEARNING OBJECTIVES
By the end of this unit you should:
✓ be able to describe rotary, linear, reciprocating and oscillating motion
✓ be able to describe methods of changing from one type of motion to another using a crank and slider, cam mechanisms and screw thread methods.

TYPES OF MOTION
There are four basic types of motion:

Rotary motion
Rotary motion follows a circular path, such as the blades of a fan, bicycle wheels, pulleys, gears and the hands of a clock.

Linear motion
Linear motion follows a straight-line path.

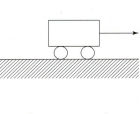

Reciprocating motion
Reciprocating motion is a back and forth motion in a straight line, such as the motion of a yo-yo or the piston in an engine.

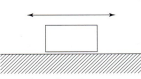

Oscillating motion
Oscillating motion is where a device swings back and forth, such as the motion of a child on a swing or the pendulum of a clock.

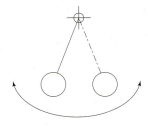

CONVERTING ROTARY MOTION TO LINEAR MOTION
Converting rotary motion to another form of motion is a common feature of mechanisms. In many cases, rotary motion is the main input source. Rotary to linear motion can be achieved using:
- a crank and slider mechanism
- cams
- screw-thread mechanisms.

Crank and slider mechanisms

Crank and slider mechanisms are frequently used to convert rotary to linear motion. The mechanism consists of three main parts: (i) a crank which is used to drive a slider; (ii) a slider (which in most case slides inside a guide); (iii) a connecting rod which joins the crank to the slider.

Slider and crank mechanisms give one type of motion only, a constant linear back and forth motion called simple harmonic motion. Crank and slider mechanisms can be used in one of two ways:

i) where the crank rotates and drives the slider, for example, operating sewing machine needles, producing the motion of a mechanical saw and driving steam engine wheels

ii) where power is applied to the slider which turns the crank via the connecting rod. This principle is used in the internal combustion engines where pistons slide backwards and forwards to drive a crankshaft. This in turn can be used to drive the wheels of a vehicle.

For crank and slider mechanisms, the slider stroke length is equal to twice the length of the crank (or twice the radius of a disc or solid wheel if used as part of the slider and crank mechanism).

Length of slider stroke = 2 × length of the crank = 2 R where R = length of the crank.

For example, if a slider and crank mechanism has a crank length of 105 mm then the stroke length = 2 × R = 2 × 105 mm = 210 mm.

The frequency of the slider (the number of cycles it makes for one revolution of the crank) is related to the rotational speed of the crank and also to the stroke length of the slider.

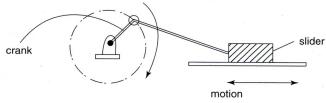

△ A crank and slider motion

Example

A crank attached to a slider and crank mechanism has a length of 200 mm. If the rotation speed of the crank is 120 rpm what will be the distance covered by the slider per second?

The stroke length of the slider = 2 × R = 2 × 0·2 = 0·4 m

The rotational speed of the crank in metres per second
= $\frac{120}{60}$ rpm = 2 revs/s

The total distance covered by the slider in one second
= 2 × 0·4 = 0·8 metres

SKILLS ACTIVITY

Using card, paper and clips, design and make a simple flat card model of a slider and crank mechanism which can be used to demonstrate a slider and crank mechanism with a slider stroke of 120 mm. Describe the motion of the slider when the crank is rotated through one revolution.

Eccentrics

An eccentric is a circular disc fixed to a shaft, but with its centre offset from the axle. Eccentrics are often used in engines and pumps to convert the rotary motion of the shaft to the linear motion required by other parts of a mechanism. In a similar way to a slider and crank, it creates linear motion for the part it is driving.

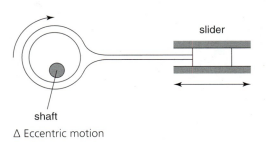

△ Eccentric motion

Cams

A **cam** is another part of a mechanism designed to change rotary motion to linear motion. It is made up of three parts: the cam, the slide and the follower. The cam shaft rotates and the follower is a rod that rests on the cam. The follower can either move up or down or dwell (stay still).

Unlike the crank and slider mechanism, the output motion of a cam can differ depending on the shape of the cam profile. Cams can be found in toys, household appliances, means of transport (cars, lorries, bicycles) and industrial equipment, such as sewing machines.

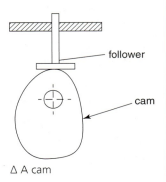

△ A cam

Cams are useful in mechanisms because they have a small number parts and can be designed to transmit different patterns of motion, depending on the shape or profile of the cam. Cams tend to be reliable - they can work at high speeds and can operate relatively quietly. There are several types of cam.

- Plate cams are common. They have a specific profile edge depending on the type of motion required.
- Profile cams are produced using flat materials such as sheet metal, plastics or wood. In many cases, it is necessary to machine the profile of the cam to a high precision so that the cam can operate effectively.
- Barrel cams have grooves machined into the cam to provide a track for the cam follower to follow.

The type of motion depends on the shape of the cam. The cam performs three basic actions:

1. a rise where the cam follower moves up

2. a fall where the cam follower drops down

3. a dwell where there is no rise or fall movement.

The follower of a plate cam rests against the cam profile and follows the shape of the cam as it rotates. Followers can be flat, roller, knife edge or they can be levers which amplify the movement caused by the cam.

The shape of the cam determines what kind of motion is produced. These shapes include pear-shaped cams, circular or eccentric cams, snail cams, and cams with multiple rise and falls.

In a pear-shaped cam, for example, the follower will move backwards and forwards in relation to the shape of the outside surface of the cam.

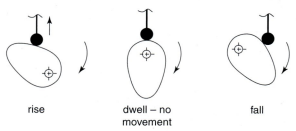

△ Motions of different types of cams

The eccentric is a disc with the centre of rotation moved away from the centre of the circle. The snail cam can only operate in one direction. If it is turned the other way it will lock against the follower. The snail cam produces a sudden drop when the follower reaches the vertical portion of the profile. If more than one movement is required for each revolution of the cam, a profile such as an ellipse, a square or a hexagon can be used to produce two, four or six lifts respectively.

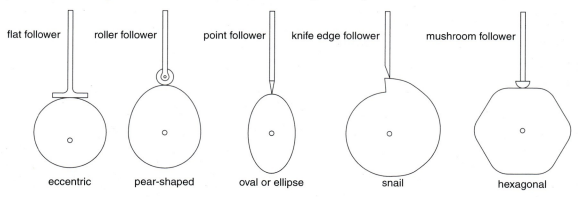

Displacement diagrams show the rise, fall or dwell characteristics of a cam. They relate the angular position of the cam to the displacement (movement) of the cam follower. The diagram below shows a simple displacement diagram for a cam and follower system. The diagram shows that the cam follower moves with an initial constant velocity rise, then a dwell where the follower does not rise or fall and finally a constant velocity fall to the start position.

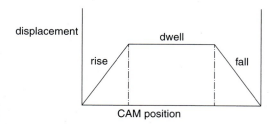

SKILLS ACTIVITY

A cam has been designed to provide a constant velocity rise, a dwell period and constant velocity fall. The rise, dwell and fall characteristics of the cam are as follows. Draw a displacement diagram for the operation.

- constant velocity rise of 10 mm for 0·25 of the cam revolution
- a dwell for 0·5 of the cam revolution
- a fall of 10 mm for the last 0·25 period of the cam revolution

Screw threads and the screw jack

Screw threads are often used to convert rotary to linear motion. When the screw thread rotates, a collar (nut) can move relative to the screw thread rotation. The smaller the pitch, which is the distance between the crests of the thread, the greater the mechanical advantage. Examples include screw jacks and turn-buckles.

A **screw jack** is a type of lifting device operated by turning a screw. The jack can be used to jack up cars or parts of buildings that are under construction.

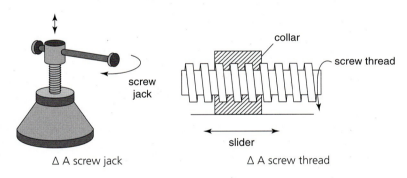

△ A screw jack   △ A screw thread

The input power for the jack is applied using a handle which acts as a lever. In the simplest case (that is, using a single start thread), one revolution of the handle results in the jack moving a distance equal to the pitch of the screw thread. This means:

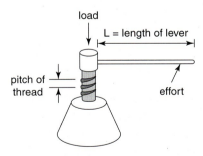

$$\text{velocity ratio} = \frac{\text{distance moved by the effort}}{\text{distance moved by the load}}$$

$$\text{velocity ratio} = \frac{2\pi L}{P}$$

where L = length of the screw jack lever
P = the pitch of the screw thread

Example

A screw jack used to lift a car has a screw thread with a pitch of 4 mm. If the length of the lever on the screw jack is 250 mm, what will the velocity ratio of the jack be?

$$\text{velocity ratio} = \frac{2\pi L}{p}$$

$$\text{velocity ratio} = \frac{2\pi \times 250}{p}$$

$$\text{velocity ratio} = \frac{2\pi \times 250}{4} = 392 \cdot 7$$

SKILLS ACTIVITY

A screw jack has a 300 mm lever and a screw thread with a 3 mm pitch. Determine the velocity ratio of the jack. If the jack is able to lift a mass of 80 kg by applying an effort of 50 N, what is the mechanical advantage of the system?

Wheel and axle

Wheel and axles are used for applications such car steering wheel mechanisms, wheel and axle parts of cars and locomotives, and other types of transport. Wheel and axle also describes the operation of screwdrivers and door knobs. To operate a wheel and axle, an effort is applied to a large diameter wheel which transmits a load to a smaller diameter wheel, shown in the images below.

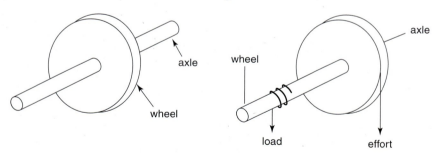

The mechanical advantage and velocity ratio of a wheel and axle system can be determined by:

$$\text{Velocity ratio} = \frac{\text{distance moved by the effort}}{\text{distance moved by the load}}$$

The distance moved by the effort during one turn of the wheel is equal to $\pi \times$ diameter of the wheel.

The distance moved by the load during one turn of the axle is then equal to $\pi \times$ diameter of the axle.

$$\text{Velocity ratio} = \frac{\text{diameter of the wheel}}{\text{diameter of the axle}} \quad \text{(note - the } \pi \text{ values cancel out)}$$

The mechanical advantage of a wheel an axle works out to be the same value as the velocity ratio. This is because in a similar way the ratio is related to the relative diameters of the wheel and axle.

$$\text{Mechanical advantage} = \frac{\text{diameter of wheel}}{\text{diameter of axle}}$$

Example

A steering wheel on a car has a diameter of 16 cm and is attached to a steering column which has a diameter of 4cm. If the driver steers the wheel with a force of 10 N what will be the force exerted on the steering column shaft?

$$\text{Mechanical advantage} = \frac{\text{diameter of the wheel}}{\text{diameter of the axle}}$$

Mechanical advantage = 16 cm ÷ 4 cm

Mechanical advantage = 4

The force exerted on the steering column shaft is therefore 10 N × 4 = 40 N.

SKILLS ACTIVITY

(i) A car has a wheel with a diameter of 600 mm. The axle that transmits power to the wheel has a diameter of 25 mm. What is the mechanical advantage of the car's wheel?

(ii) The back wheel of a bicycle has a diameter of 700 mm. If the back wheel axle has a diameter of 9 mm, determine the mechanical advantage.

KNOWLEDGE CHECK

1. Describe the following types of motion: rotary, linear, reciprocating and oscillating. (4)
2. What are the main parts of a cam system? (2)
3. Draw a crank and slider mechanism and explain how the linear motion of the slider is achieved. (4)

KEY TERMS

cam: a specially shaped part attached to a rotating shaft that imparts linear motion to a cam follower

crank and slider: a mechanism used to change rotary motion to linear motion, or vice versa

linear motion: motion that follows a straight line

oscillating motion: motion backwards and forwards or from side to side

reciprocating motion: a repetitive motion backwards and forwards or up and down

rotary motion: motion in a circular path

screw jack: a device designed to lift a load, operated by turning a leadscrew

wheel and axle: a simple machine part that consists of a wheel attached to an axle and which is used to lift loads or transmit power from one part of a mechanism to another

4·3·3 Transmission of motion

LEARNING OBJECTIVES

By the end of this unit you should:
✓ understand how gears are used to transmit motion between shafts which are close together
✓ understand how belt drives are used to transmit motion between shafts which are relatively far apart
✓ be able to describe different types of gears; for example, spur gears, helical gears, bevel gears, worm and worm wheels, and rack and pinions
✓ understand the difference between simple gear trains and compound gear trains
✓ be able to describe different types of belt systems
✓ be able to calculate the speed ratios for gear train and belt drives systems.

There are a number of ways motion is transmitted between rotating shafts and other parts of mechanisms. These include gear trains, **pulley** systems, sprocket and chains, and rack and pinion devices. Gears are used to transmit rotary motion between shafts which are relatively close together. Pulley systems and sprocket and chains are used to convert motion between shafts which are relatively far apart. A rack and pinion device is a method of changing rotary motion into linear motion or visa-versa.

GEARS AND GEAR SYSTEMS

Gears are key components of many mechanisms. Gears can be used to increase or decrease the speeds of shafts or to change the direction of motion. They are also used in gear boxes to increase or decrease the power output of machines such as car engines or bicycles. A gear has a specific number of teeth, which mesh with the teeth of other gears.

A **gear system** comprising of two gears meshed is called a simple gear train. Two gears fixed together on a single axle form a compound gear train; this allows large speed reductions in a small space.

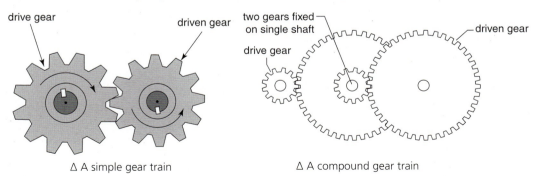

△ A simple gear train   △ A compound gear train

In gear systems, the gear which is driving the gear train is called the

driver gear. The driver gear is often attached to a shaft being driven by an electric motor, generator or engine. The gear meshed with the driver gear is called the **driven gear**.

When two gears mesh they rotate in opposite directions. However, an idler gear can be used so the gears rotate in the same direction, shown below.

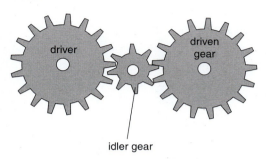

The velocity ratio (often called gear ratio) of the two gears is determined from the relationship:

$$\text{Velocity ratio} = \frac{\text{number of teeth on the driven gear}}{\text{number of teeth on the driver gear}}$$

Example

A simple gear train has a driver gear rotating at 200 rpm. If the number of teeth on the driver gear are 40 and the number of teeth on the driven gear are 80, what will be the output speed of the driven gear?

$$\text{Velocity ratio} = \frac{80}{40} = 2$$

The output speed will then be 200 ÷ 2 = 100 rpm

SKILLS ACTIVITY

What will be the output speeds of the gear systems shown in the table below?

Input speed (rpm)	Number of teeth on the driver gear (T)	Number of teeth on the driven gear (T)	Output speed (rpm)
500	20	60	
1500	35	105	
700	90	30	

VELOCITY RATIO OF COMPOUND GEAR SYSTEMS

The velocity ratio of a compound gear train can be determined by multiplying each individual gear ratio together.

For example, for the compound gear train shown in the diagram, the velocity (gear) ratio becomes:

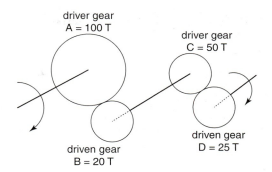

$$\frac{\text{driven gear (B)}}{\text{driving gear (A)}} \times \frac{\text{driven gear (D)}}{\text{driving gear (C)}}$$

$$= \frac{20}{100} \times \frac{25}{50} = \frac{500}{5000} = \frac{1}{10}$$

SKILLS ACTIVITY

The diagram shows four gears meshing together. Determine (i) the velocity ratio of the gear train and (ii) the output speed of gear D if the input speed of gear A is 130 rpm. If gear A is changed to a gear with 60 teeth and gear D to 30 teeth, what will the new output speed be?

GEAR SYSTEMS AND TORQUE

A gear system will also transmit a turning force to the shaft it is attached to. A turning force of this nature is known as torque and its units are Nm. There is a close relationship between the torque produced by two gears meshing and the velocity ratio of the system:

$$\text{Velocity ratio} = \frac{\text{torque at output}}{\text{torque at input}}$$

Example

A two gear system has a driver gear with 120 teeth and a driven gear with 40 teeth. If the input gear produces a torque of 35 Nm, what will be the output torque produced by the output gear?

$$\text{Velocity ratio} = \frac{\text{number of teeth on the driven gear}}{\text{number of teeth on the driver gear}} = \frac{40}{120} = 3$$

$$\text{Velocity ratio} = \frac{\text{output torque}}{\text{input torque}}$$

Output torque = velocity ratio × input torque = 3 × 35 = 105 Nm.

SKILLS ACTIVITY

(i) A gear system has a velocity ratio of 4. If the torque transmitted by the input gear is 40 Nm what will be the torque produced by the output gear?

(ii) For a simple gear train, the driver gear produces a torque of 70 Nm allowing the driven gear to produce a torque of 210 Nm. What velocity ratio is necessary to produce this? If the driver gear has 30 teeth, what number of teeth should be selected for the driven gear to give this velocity ratio?

TYPES OF GEARS

There are different types of gears for different purposes. The main gear types are:

Spur gears

Spur gears are the most common types of gears. They have straight teeth and are typically made from steel or plastics such as nylon. A problem with spur gears, especially when used at high speeds, is that they are noisy due to the vibration occurring between the teeth when meshing.

Helical gears

Helical gears are used to reduce the noise and vibration associated with spur gears. They can also withstand high loads. This is because when the gears mesh the loads are distributed over several teeth rather than just one, as with spur gears.

Bevel gears

Bevel gears are often used to change the axis of rotation of shafts through 90°, such as in a hand-food mixer.

Worm gear systems

Worm and worm-wheel systems tend to be used when large speed reductions between shafts are required in a small space. In the system, the worm is the driver gear and the worm wheel or pinion is the driven gear. The worm-wheel is locked in place and cannot turn the worm gear. They are commonly used in hoist mechanisms.

The velocity (gear) ratio of a worm gear = $\dfrac{\text{number of teeth on the worm-wheel}}{\text{number of teeth on the worm}}$

However, because the worm acts as a single-toothed gear for each revolution, the gear ratio becomes:

Velocity ratio of a worm gear = $\dfrac{\text{number of teeth on the worm-wheel}}{1}$

Example 1

A worm and worm-wheel system has a worm-wheel with 40 teeth. What will be the gear ratio?

Gear ratio = 40:1

Example 2

For a worm-wheel system where the worm-wheel has 50T, determine the gear ratio. What will be the output speed if the worm-wheel has an input speed of 200 rpm?

The gear ratio = 50:1

This means that the worm gear reduces the output speed by fifty times. Therefore, the output speed is 4 rpm.

SKILLS ACTIVITY

(i) If a worm-wheel has 60T what is the gear ratio? What will be the output speed for input speeds of (i) 800 rpm and (ii) 120 rpm?

(ii) Using a sketch, describe how a worm and worm-wheel could be used as part of a hoist system.

Rack and pinion systems

Rack and pinion gears are used to convert linear to rotary motion or vice versa. The spur gear is called the pinion and the straight gear is called the rack. When the spur gear rotates it will cause the rack to move in a linear direction, such as below.

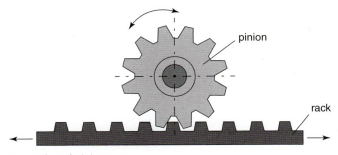

△ A rack and pinion gear

The relationship for a rack and pinion is as follows:

the distance moved by the rack per turn = $\dfrac{\text{number of teeth on the pinion}}{\text{number of teeth on the rack}}$

For instance, if a rack with 200 teeth per metre is moved by a pinion with 20 teeth, then the distance the rack moves (if the pinion turns through 4 revolutions) is $\dfrac{20}{200} \times 5$ turns = 0·5 m.

Ratchet and pawl

A ratchet and pawl allows rotation in one direction only, preventing rotation in the opposite direction. They are commonly used for controlling sports stadium turnstiles, ratchet spanners and raising and lowering lock gates.

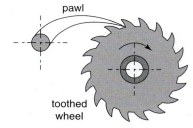

△ A ratchet and pawl

BELT DRIVES AND SPROCKET AND CHAINS

Belt drives are used to link two rotating shafts which are relatively far apart. In most cases the shafts are parallel to each other. In a similar way to gear systems, belt drive pulleys are either driver or driven pulleys. The driver is generally attached to a motor or other power source. The driver then drives the driven pulley using a connecting belt.

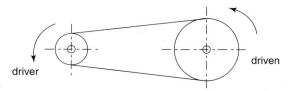

△ A belt drive with pulleys rotating in the same direction.

When a belt is used without crossing, the pulleys will rotate in the same direction. If the belt is crossed then the pulleys will rotate in opposite directions. This is a common method of changing the direction of rotating shafts.

Pulley systems tend to be cheaper than gear systems. However, a problem with pulleys is that the belts can slip and stretch over time. Most belt drive systems, therefore, need a method of adjusting the belt tension.

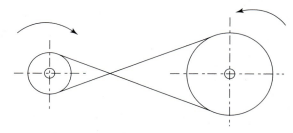

△ A crossed belt drive with pulleys rotating in the opposite direction.

Belt drives are classified by the cross-sectional shape of the belt and include:
- flat belts
- round belts
- vee belts
- toothed belts.

Velocity ratio and torque calculations for belt drive examples

A motor drives a pulley system with a torque of 150 Nm. If the diameter of the driver pulley is 50 mm and the diameter of the driven pulley is 100 mm, what is the torque produced by the driven pulley?

Velocity ratio of the pulley system = $\dfrac{\text{diameter of driven pulley}}{\text{diameter of the driver pulley}} = \dfrac{100 \text{ mm}}{50 \text{ mm}} = 2 \cdot 0$

Velocity ratio = $\dfrac{\text{output torque}}{\text{input torque}}$

Output torque = input torque × VR = 150 Nm × 2·0 = 300 Nm

SKILLS ACTIVITY

(i) A belt drive has a driver pulley of 220 mm diameter and runs at 120 rpm. If the driven pulley has a diameter of 100 mm, what is its speed? What would be the output speed if the belt drive is 90% efficient due to slippage between the belt and pulleys?

(ii) Complete the table below and determine the output torque in each case.

Driver pulley diameter (mm)	Driven pulley diameter (mm)	Velocity ratio	Input torque (Nm)	Output torque (Nm)
300	60		40	
100		1:5	60	
60	20		100	

Flat belts

Flat belts are a common type of belt drive. However, because this type of belt tends to 'climb' off the face of the pulley when used, the pulley surface tends to have a slight convex shape. This keeps the belt central when it is rotating. Flat belts are commonly used to drive parts of textile machinery. They are found in household appliances such as vacuum cleaners and are used in small electrical generators.

Round belts

Round belts have a circular cross-section and tend to run in pulleys with a vee-groove. When the belt is rotating, the sides of the belt 'squeeze' out and press against the sides of the vee. Round belts are used for low power transmission applications, such as driving the needle in sewing machines and small electrical fans.

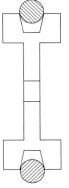

Vee-belts

Vee-belts are used where high-transmission forces are required. The vee-shape of the belt fits into the vee-groove of the pulley. Vee belts are used widely in industry where high loads need to be driven at high speeds without the belt slipping. Typical uses are driving lathes, milling machines, industrial presses and industrial conveyors.

Toothed belts

Toothed belts have teeth which match up with teeth in a toothed pulley. Toothed belts are used to give a positive drive between two pulleys. This is used to reduce belt slippage and give a synchronised drive between two shafts of a mechanism. They are commonly used for timing applications, such as in car engines and in computer printer equipment.

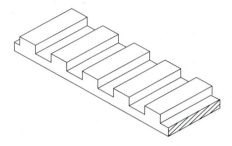

SPROCKET AND CHAIN DRIVES

Sprocket and chains are commonly used for driving the wheels of bicycles and motorbikes, and for driving the timing belts for some motor car engines. The sprocket and chain allows power to be transmitted from one shaft to another without slippage. This means that relatively high power can be transmitted from one shaft to another. The sprocket is the toothed wheel in which the chain engages.

△ A sprocket and chain

SYSTEMS FOR MAINTAINING TENSION IN BELT DRIVES AND CHAIN DRIVES

A problem with many belt drive systems is that the belt can stretch over time. This results in loss of transmission between the belt and the pulley it is driving. To help reduce the effect of overstretching, belt drives include a method to adjust the belt tension when needed. It is important to ensure that belts are adjusted correctly when tensioned. If they are tightened too much, then the resulting tensioning force can damage the belt drive bearings when running. To maintain the tension of a chain drive, a link (or links) in the chain can be taken out to shorten the chain. There are a range of adjustment methods commonly used for belt drives; a selection of these are shown below and on the next page.

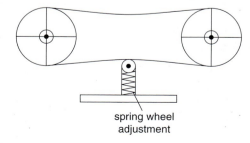
spring wheel adjustment

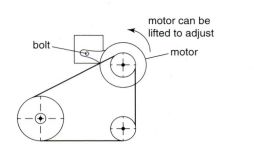
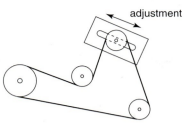

PNEUMATIC AND HYDRAULIC SYSTEMS   EXTENDED

Pneumatic systems make use of compressed air to operate mechanisms and equipment. **Hydraulic systems** use a fluid such as hydraulic oil. Using oil means that hydraulics tend to be used for higher load applications, where positive movement and no cushioning is required; for example, in hydraulic shock absorbers for automobiles and motorcycles as well as landing gear for aircraft. It is also used for more precise applications such as positioning computer-controlled machinery (CNC machinery).

Pneumatics are widely used in food, medical and textile applications where minimal contamination is required. Hydraulic systems are not usually used in such cases because of the possibility of leaks from faulty joints, valves and pipes. Pneumatic systems usually operate faster than hydraulic systems. This is because oil can be 'sluggish' when working in the system.

Pneumatics use compressed air to operate mechanisms. The compressed air is produced using a piece of equipment called a compressor. A compressor compresses ordinary atmospheric air to the higher pressure needed for the compressed air system. Pneumatic systems are typically made from joining four types of equipment:

- a compressor used to supply the compressed air to the system
- the lines (pipes) used to connect equipment together
- valves used to control and direct the compressed air to equipment
- cylinders that provide linear movement when operated by the compressed air.

Cylinders

Compressed air cylinders are used to provide linear motion when operated by a pneumatic valve. Two types of cylinders are (i) single acting cylinders and (ii) double acting cylinders.

A single acting cylinder has one air supply. When air is supplied to the cylinder, the piston inside the cylinder is pushed out, or outstrokes. When the air supply is stopped, a spring inside the cylinder returns the piston, or instrokes, to its start position. This type of action is useful for applications such as thread cutting operations on sewing machines, where a knife is attached to the end of the piston to cut the thread when operated, or for packaging systems where the cylinder is used to move packages from one part of the system to another.

A double acting cylinder has two air supply connections. One is used to push the piston in one direction, the other is used to push the piston in the opposite direction. There is no spring in the cylinder, as is the case with the single acting cylinder.

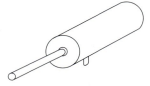

△ Single-acting cylinder

A double acting cylinder works in the following way: when compressed air goes through port A it pushes the piston outwards. The air in front of the piston will then be exhausted through port B. To reverse the process, air is directed to port B which has the effect of moving the piston towards its original start position. The air is now exhausted through port A. A compressed air valve (a type of switch) is used to control the compressed air flow to each of the A and B ports.

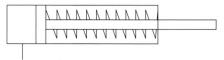

△ Inside view of single-acting cylinder

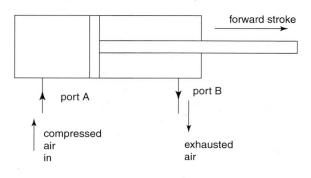
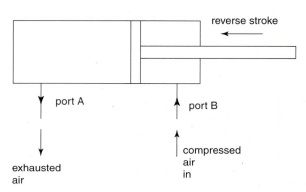

In both cases, the cylinders are operated by pneumatic valves. The valves direct air to the cylinders. A valve called a $\frac{3}{2}$ valve is used to operate a single acting cylinder. A $\frac{5}{2}$ valve is used to operate double acting cylinders.

INTEGRATION OF ELECTRONIC AND PNEUMATIC SYSTEMS

EXTENDED

Many valves are now integrated with electronic sensors and electronic switching devices, so that they can be operated by microprocessors and other electronic circuitry. This allows a more integrated approach to the control of pneumatic systems. There are a number of advantages of using such an approach, including the principle that valves can be operated automatically with no human intervention. Valves can also be operated by control systems which are a long way from the compressed air system it is controlling. Solenoid valves are commonly operated by electronic circuitry. These, for example, can be 3/2 or 5/2 valves. Over the last few years, there has been a trend to build miniature valves which can allow compressed circuitry to be packed into smaller spaces for many control applications.

△ An electronic valve

SELECTING MECHANISMS FOR DESIGN APPLICATIONS

When selecting parts for a mechanism, a number of factors should be considered. These include:

1. What type of motion is required for the mechanism part or parts (for example, linear motion or oscillating motion)?
2. The speed the mechanism parts need in order to run
3. How much space is available for the mechanism parts?
4. The loads that need to be carried and transmitted by the mechanism parts
5. The types of materials available for the mechanism parts, for example, plastics or metals
6. The environment the mechanism will run in
7. The cost of the product
8. Maintenance requirements
9. The torque required by mechanism parts
10. Whether vibration is an important consideration

KNOWLEDGE CHECK

1. Give two advantages of using gears rather than belt drives. (2)
2. Explain the difference between a simple gear train and a compound gear train. (2)
3. What is meant by a positive belt drive? (2)
4. What advantages does pneumatics have over hydraulics when used in food applications? (2)

KEY TERMS

driven gear or pulley: a gear or pulley driven by a driver gear or pulley

driver gear or pulley: a gear or pulley used to drive the system it is connected to

gear system: a system of toothed wheels which connect to transmit motion between shafts

hydraulic system: a system which operates using oil or water

pneumatic system: a system which operates using compressed air

pulley: a wheel with a grooved rim in which a belt runs

4.3.4 Energy EXTENDED

LEARNING OBJECTIVES

By the end of this unit you should:
- ✓ understand that energy sources include mains electricity, batteries, solar panels and generators
- ✓ be able to explain the uses of rechargeable batteries
- ✓ understand that photovoltaic (PV) solar cells convert light energy to electricity
- ✓ understand that a generator converts mechanical energy to electrical energy.

MAINS SUPPLIES

Mains supplies are common for household and industrial applications. For safety reasons, high voltages are often transformed to lower voltages so that equipment can be run off a mains supply. For example, transformers are used to charge up and power devices such as mobile phones and computers. Mains supplies are dangerous and precautions have to be taken to eliminate any risk of electrocution.

△ Mains supplies are available in most buildings in order to power common electrical appliances.

Most countries will use some form of residual current device (RCD), sometimes known as an earth leakage circuit breaker (ELCB). These devices will disconnect the supply quickly if there is a fault, in a fraction of the time that a fuse or circuit breaker would take. The RCD will measure the difference between the current going into an appliance and the current coming out of the appliance. Any differences are generally caused by a fault in the appliance. The supply is then cut off immediately, reducing the risk of death or injury to a user and also reducing the risk of fire.

BATTERIES

Batteries can be relatively large, for example, car batteries. However, many batteries are considerably smaller when used for devices such as wrist watches, mobile phones, remote control devices, and cordless devices such as GPS systems and electronic cameras.

△ AA batteries power common portable devices such as remote controls.

Batteries can either be non-rechargeable or rechargeable. Non-rechargeable batteries only last for a certain period of time, and once they are discharged they have to be disposed of. Rechargeable types can be recharged using a battery charger, which is often part of the device or equipment the battery is powering.

Many of the materials used in battery construction are either toxic or hazardous. Lead and sulphuric acid are both used in car batteries. Hearing aid batteries may contain mercury and Nickel

Cadmium (NiCd). Rechargeable batteries also contain hazardous waste. The main danger occurs when the casing is damaged and the chemicals in the battery are exposed.

SOLAR CELLS

Solar cells (also known as photovoltaic cells) convert light energy into electrical energy. They are often used to power devices such as calculators, which have the solar cell as their power source, with a back-up battery for use when there is no sunlight. Solar cells are also used for powering remote equipment such as satellites and space stations, or traffic signs in the countryside.

△ Solar cells are often the power source of calculators.

GENERATORS

A **generator** is a device used to convert mechanical energy to electrical energy. A dynamo, which can be used to generate electricity from a bicycle wheel, is a type of direct current generator. A turbine is a type of generator which uses heat energy, often in the form of steam, to drive a generator and generate the electrical energy on a large scale. Alternatively, a wind generator converts the linear force of wind to the rotary motion of the wind turbine blades to power the turbine.

△ This dynamo is a direct current generator.

ENERGY COSTING

The amount of energy used by electrical appliances can be determined by the power rating of the appliance and the amount of time it is on for. An appliance will have a particular power rating depending on the amount of current it draws from the electrical supply. The higher the power rating, the more it will cost to run over a period of time. The standard unit of mains electricity in many countries is the kilowatt-hour (kWh). This is defined as the power in kilowatts used by an appliance multiplied by the time it is switched on.

That is: kWh = power (kW) × time (hours)

Example

A washing machine rated at 1·5 kW runs for 30 minutes. How many units of electricity has been consumed by the appliance. If it costs $0.5 per unit, what will be the cost of running the appliance?

kWh = 1·5 kW × 0·5 = 0·75 kWh

Therefore the total cost of running the machine for 30 minutes = 0·75 × $0·5 = $0·37

SKILLS ACTIVITY

For each of the cases given, determine the cost of running each appliance.

Appliance	Power rating	Time the appliance is running	Cost per unit
Electrical kettle	1.8 kW	5 minutes	$0.8
Television	50 W	2 hours	$0.5
Steam iron	2000 W	1.5 hours	$1.6

SELECTING ENERGY SOURCES

When selecting energy sources for systems applications, it is important to work towards using the source in the most efficient way. There are a number of ways to reduce the energy consumption of products. These include using insulation to reduce the overall energy losses, making use of energy efficient components such as LEDs, and closely matching the energy supply to the voltage requirements of product.

When operating systems, working towards reducing energy use is an important strategy in energy efficient and sustainable design. Reducing energy consumption in all areas of systems and control work helps to reduce the energy that is needed to work, transport and maintain such systems.

Energy efficiency can be achieved by using rechargeable batteries for certain applications, by using renewable energy products (such as solar panels) and by using mechanism parts (such as motors) with low energy requirements. Energy can also be conserved by the design of efficient mechanisms and other systems. This can be done by using the least number of parts as possible for the design and by making good use of bearings and lubrication to reduce energy loss through heat, wear and friction. It is also important to consider using materials which are energy efficient to manufacture and which will reduce the energy losses when a mechanism is operating.

> **KEY TERMS**
>
> **battery:** an electrical device that can store energy
>
> **generator:** a machine that converts mechanical energy into electrical energy
>
> **kilowatt-hour:** a measure of electrical energy consumption equal to the kilowatts used by an electrical appliance for every hour it is switched on
>
> **mains supply:** a general purpose electrical supply for buildings, supplied through cables
>
> **solar cell:** a device which converts light to electrical energy

4·3·5 Bearings and lubrication

LEARNING OBJECTIVES

By the end of this unit you should:
✓ understand the purpose of lubrication
✓ be able to describe the use of plain bearings, ball bearings, roller bearings and thrust bearings.

BEARINGS

A **bearing** is a machine component which reduces the friction between a rotating shaft and the part of machine it is rotating in. Bearings are found in many different kinds of machines and equipment. A bearing allows the shaft to rotate freely in the machine part, and in doing so prevents friction between the shaft and its housing. Without a bearing the shaft would seize up (lock) and the machine would not be able to operate under such conditions. Common types of bearings include:

△ Ball bearings come in lots of different sizes.

Plain bearings

A plain bearing is a simple type of bearing that has a bearing surface with no ball or roller elements. Plain bearings are often made from phosphor bronze or brass, which are softer materials than the axle that will run in them.

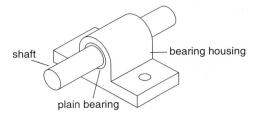

Ball bearings

A ball bearing uses balls to separate two bearing races. The outer race is generally locked onto the shaft and remains stationary when the shaft is rotating. The inner race can rotate with the shaft. Ball bearings can be sealed for life, meaning that there is no maintenance.

Roller bearings

A roller bearing uses rollers that revolve between the two bearing races. Roller bearings can take higher loads than ball bearings.

Thrust bearings

A thrust bearing is designed to withstand axial loads (which are also known as thrust loads). They are designed to accommodate loads in the axial direction of the shaft they are attached to. The most common types are ball thrust bearings and roller thrust bearings. Roller thrust bearings tend to be

used where there is a need to withstand high axial forces, such as the axial loads on ship propeller shafts and car transmission applications.

Air bearings

Air bearings use a thin layer of pressurised air to support a load. The bearing has no solid-to-solid contact when it is operating. There is no metal-to-metal contact, as is also the case for roller bearings or ball bearings. Air bearings do not suffer from heat or wear due to friction. The diagram below shows the bearing works by high pressure air being fed into the gap between the rotating shaft or other component, which then allows the shaft to rotate.

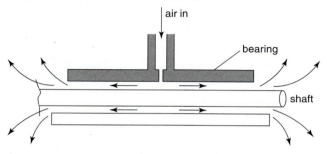

Magnetic bearings

A magnetic bearing uses a strong magnetic field to support a load. This prevents physical contact between the two bearing surfaces. The advantages of using magnetic bearings are that there is no mechanical wear and there is no need for oil or grease seals. Very high speeds can be achieved when using magnetic bearings. Magnetic bearings are often used in blood pumps and other high-precision engineering applications.

SELECTION AND SUITABILITY OF BEARINGS FOR SPECIFIC CONDITIONS

The selection of bearings for specific design applications depends on a number of factors. These include: the type of load the bearing needs to withstand, the type of environment it will be used in, its operating speed and temperature range. For example, plain bearings tend to be used for relatively low-speed applications and require good **lubrication** to stop them seizing up when operating. Roller and ball bearings, on the other hand, tend to be used for higher-speed and higher-load applications.

The type of working environment is also a factor in bearing selection. Bearings which are designed to work in dusty environments need to be protected with covers, rubber seals and other types of enclosures. Enclosure methods are needed to protect the bearing from damage from dust, metallic fragments, grit or other particles. Seals for bearings need to be designed to stop contamination materials getting into the bearing. Seals also act to keep the oil, grease or other lubricant materials inside the bearing area.

A key selection factor is the type of load the bearing is designed to withstand. Many bearings, for example, radial roller bearings or ball bearings, are selected as they are located on a shaft in such a way as to be able to resist radial loads. However, thrust bearings are designed to withstand axial loads. These loads are in line with the main axis of the shaft and therefore are required to withstand a 'thrusting' type of load.

The operating temperature of the bearing is also an important factor to consider when matching a bearing type to a particular application. Many plain bearings, for example, are designed to operate at temperatures between −30 C° and +50 C°. Roller bearings and ball bearings can withstand a higher temperature range.

Bearings have a limited service life and should be checked and maintained on a regular basis. Inspection often includes checking the state of the lubricant, checking bearings for wear and tear, and judging whether the bearing needs to be changed. When designing machine parts, it is good practice to consider the ease at which the bearing parts can be inspected and how easy it is to maintain and change the bearing if required.

LUBRICATION

Lubrication is used to prevent wear between two surfaces. When two surfaces rub together they produce friction and this can cause the surfaces to stick together and also wear them away over time.

Lubrication helps prevent metal-to-metal contact by producing a thin layer to push the two contacting surfaces apart. The main lubricants are oil and grease. Oil is good because it is thin enough to easily get between closely fitting parts such as gear teeth. It is also good at taking away the heat that is generated by friction. Oil needs to be changed regularly because small particles that break away from the sliding surfaces can contaminate the oil causing further damage.

PTFE bearings

PTFE (polytetrafluoroethylene) is typically used as a coated layer to provide a non-stick surface. It is often used as a coating for non-stick pans and other cooking products. PTFE can come in spray-can form for applications such as lubricating locks and clock mechanisms.

Solid film lubrication

Solid films are solid materials that reduce the friction between surfaces. These materials may be used as an additive to other lubricant materials or used in pure form. Common types include graphite, talc and molybdenum-based products. Many are chosen because they can withstand extreme operating temperatures.

SKILLS ACTIVITY

Make a list of household appliances that might need bearings to make them work properly. For each appliance identified, briefly describe why the bearings might be necessary.

KEY TERMS

bearing: a mechanical device used to reduce friction between moving parts

lubrication: the use of oil, grease or other material to reduce friction between two moving parts

4·4 Electronics

4·4·1 Basic Concepts

> **LEARNING OBJECTIVES**
> By the end of this unit you should:
> ✓ understand the terms and units for electrical current, voltage, resistance and wattage
> ✓ be able to recognise a range of circuit symbols
> ✓ understand how to use the power relationship P = V × I
> ✓ understand the difference between direct and alternating current
> ✓ be able to recognise and draw simple direct current circuits
> ✓ understand how to use meters to measure electrical current, voltage and resistance.

Electronic circuits are widely used for operating and controlling products and equipment. There are numerous examples ranging from mobile phones to more complex control systems, such as automatic flight systems in aircraft. These circuits, however complex, are made using components such as integrated circuits, discrete components, conductors, and energy sources such as batteries. The components and other parts of the circuit are usually mounted on circuit boards.

CURRENT FLOW **EXTENDED**

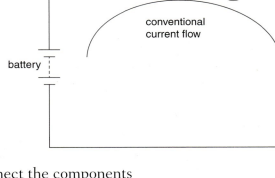

There are two types of electric flow (called **current**): direct current (DC) and alternating current (AC).

Direct current is provided by sources such as batteries, solar cells and direct current generators. The primary source of alternating current is the mains supply in homes and schools.

A simple direct current circuit is shown in the figure opposite. The circuit includes a battery, a switch and a light bulb. Wires connect the components together. The unit of current is the Amp (A). Most small circuits will have currents measured in milliamps (mA).

When the switch is closed it is assumed that the current flows from the positive battery terminal to the negative terminal. This positive to negative direction is called **conventional current flow**. In doing so it lights up the bulb. The electrical flow is in one direction only. This is why it is called direct current.

VOLTAGE

The energy supplied by batteries, generators and other sources is measured in volts (V). The higher the **voltage**, the higher the potential energy supplied to a circuit.

POTENTIAL DIFFERENCE

Potential difference is the voltage that is lost across a component when a current flows through it. The voltage loss can be measured using a voltmeter connected in parallel, such as below.

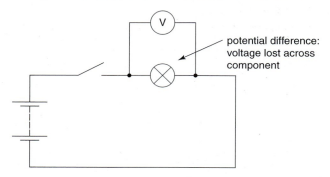

potential difference: voltage lost across component

OHM'S LAW

Resistance is measured in **ohms (Ω)**. The fundamental relationship between resistance, voltage and current in a direct current circuit is called Ohm's Law.

Ohm's Law states that Voltage = current × resistance

$$V = I \times R$$

where
- V = voltage (V)
- I = current (I)
- R = resistance (Ω)

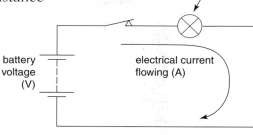

CALCULATING ELECTRICAL POWER

It is often necessary to determine the power output of electronic components and devices. For example, you might need to calculate the power dissipated by a component such as a resistor to ensure it will not become damaged by overheating if the power is too high.

Electrical power can be determined from the relationship:

power = voltage × current (I)

$$P = V \times I$$

The unit of power is the watt (W). Higher values are often expressed as kilo watts (kW) or megawatts (MW). (1kW = 1000W and 1MW = 1000kW.)

For example, if a component allows a current of 0.2 A to flow through it when its supply voltage is 10V, then the power dissipated will be:

$P = V \times I$

$P = 10 \text{ V} \times 0.2 \text{ A} = 2 \text{ W}$

In practice, electrical components have a maximum power rating (in watts), which indicates the maximum power dissipation allowed. Components must be used below this maximum rating otherwise they might overheat. A resistor, for example, limits the electrical current flowing through it. As a consequence, the current flowing through the component will produce heat which needs to be dissipated to the surroundings. A resistor designed to handle large currents will have a larger surface area than one designed to take smaller currents. Resistors commonly have values of 0.25 W, 0.5 W, 1 W and 2W. These relate to their maximum power ratings.

Example

A resistor carries a current of 0.15 A when connected to a 12 V supply. What resistor power rating should be selected?

$P = V \times I$

$P = 12 \times 0.15 = 1.8$ W

Therefore a 2 W power rating should be selected so that the resistor is not damaged when used.

SKILLS ACTIVITY

Determine the power dissipated for each case below and select an appropriate resistor value for the component.

Voltage	Current	Power dissipated (W)	Resistor power rating selected (W)
10 V	0.01 A		
12 V	0.06 A		
18 V	0.1 A		

Quantity	Unit of measurement	Symbol	Abbreviation	pico $\times 10^{-12}$	nano $\times 10^{-9}$	micro $\times 10^{-6}$	milli $\times 10^{-3}$	× 1	kilo × 10^3	mega × 10^6
Voltage	Volt	V	V			µV	mV	V	kV	MV
Electrical Power	Watt	W	W			µW	mW	W	kW	MW
Current	Amp	I	A			µA	mA	A	kA	
Capacitance	Farad	F	F	pF	nF	µF		F		
Resistance	Ohm	R	Ω					Ω	kΩ	MΩ
Frequency	Cycles per second	Hz	Hz					Hz	kHz	MHz

Converting units

$V = I \times R$ $\qquad$ $W = I \times V$

$I = V / R$ $\qquad$ $I = W / V$

$R = V / I$ $\qquad$ $V = W / I$

Resistors in series $\qquad$ $R_{total} = R_1 + R_2 + R_3 .....$

Resistors in parallel $\qquad$ $1 / R_{total} = 1 / R_1 + 1 / R_2 + 1 / R_3 ....$

Capacitors in series $\qquad$ $1 / C_{total} = 1 / C_1 + 1 / C_2 + 1 / C_3 ....$

Capacitors in parallel $\qquad$ $C_{total} = C_1 + C_2 + C_3 .....$

▽ A table of electronic symbols.

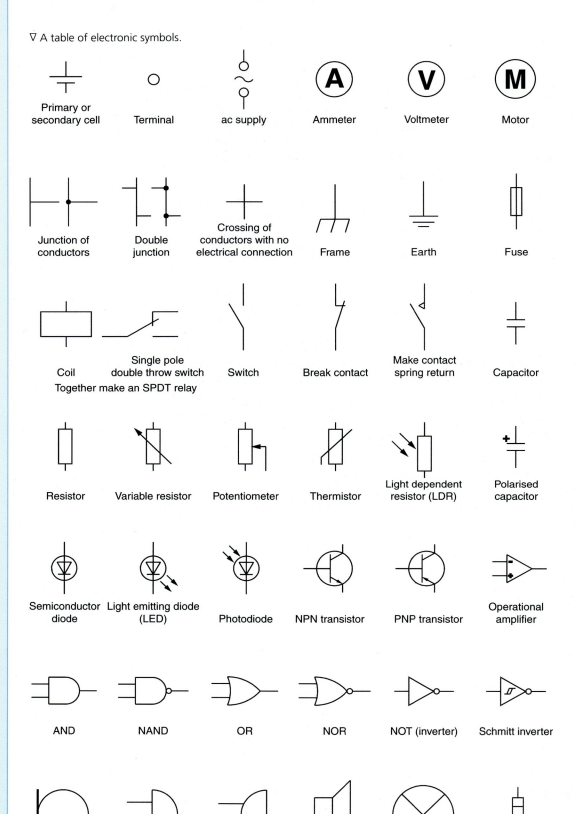

MEASURING VOLTAGE, CURRENT AND RESISTANCE USING METERS

Meters are used to help check the current, voltage or resistance values of parts of circuits. They are also used to help find faults. Ammeters and volt meters are pieces of equipment used to measure either the electrical current through components or the voltage drop across components. Multimeters are more versatile because they can measure current, voltage and resistance in one instrument.

When measuring voltage, the meter is designed to be used in parallel with the component or device being measured. To be effective, a voltmeter must have a high internal resistance.

An ammeter is used to measure the rate of an electrical current. To measure the current in a specific part of a circuit, the ammeter needs to be placed in series with the circuit. This means that the circuit needs to be broken and the ammeter placed between the break in the circuit, so that the current can be measured.

Multimeters can be analogue or digital. Analogue multimeters have a pointer and a dial on which you can read the electrical quantities. Digital multimeters have a numeric digital display, such as a liquid crystal display. Digital devices are popular because they tend to be easier to use than analogue versions, and are often more accurate. The face of a multimeter typically includes:

- a display where the values can be viewed
- buttons or a dial selection facility where you can select current, voltage or resistance measurements
- input jack sockets where you insert the leads from the device you are measuring.

◁ A digital multimeter

Multimeters can be battery operated, which means you need to check the battery before you use it to make sure that it is charged.

Care and handling of meters

It is important that you take care of meters. The following guidelines will help ensure that your meters continue to work properly.

i) Handle the instruments carefully. They are delicate and expensive.

ii) Place meters in a safe position where they cannot fall.

iii) Make sure you know how to use the instrument correctly.

iv) Make sure the instrument is not damaged before you use it. Damaged instruments may give inaccurate measurements.

v) Check the battery in the instrument (if it contains a battery).

vi) Avoid trailing test leads over the edges of work benches when the meter is in use.

vii) Carry the meter by the straps or handles provided.

SKILLS ACTIVITY

Determine the battery voltage for the circuit shown in the diagram using Ohm's law.

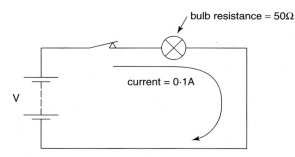

KNOWLEDGE CHECK

1. State the units of potential difference, electrical current and resistance. (3)
2. Define the term *potential difference*. (2)
3. Use Ohm's Law to determine the voltage required to produce a current of 0.25 A through a component which has a resistance of 120 Ω. (3)

KEY TERMS

electrical current: a flow of electricity through a circuit or a measure of the rate of this flow – unit of measurement: Amp (A)

potential difference: the difference in voltage across components, conductors and other sources when an electrical current flows through them – unit of measurement: Volt (V)

resistance: a material's opposition to the electrical current flowing through it – unit of measurement: Ohm (Ω)

voltage: a measure of electrical current, measured in volts – unit of measurement: Volt (V)

4.4.2 Circuit building techniques

LEARNING OBJECTIVES
By the end of this unit you should:
✓ be able to recognise good practice when designing printed circuit boards
✓ understand that strip board can be used to make simple circuits
✓ be able to recognise how prototype boards are used for building experimental circuits.

PRINTED BOARDS

Printed circuit boards (PCBs) are used because they provide a neat and reliable method of mounting components. Most electronic products incorporate a printed circuit board because they can be mass produced easily.

A printed circuit board is a copper clad board which has a thin layer of copper attached by adhesive to the base of the board. They can be either single-sided, which is usual for school projects, or double-sided for more complex circuits. Components can be inserted through drilled holes (through hole) or mounted direct to the surface of the board (surface mount).

The majority of modern electronic devices such as mobile phones (cell phones), computers, tablets, games machines and televisions use modules that have been made using surface mount components. However, components for surface mounting tend not to be used in schools because they are too small to handle and too small to **solder** using school workshop equipment.

Many low-cost computer software programs are now available to help with the circuit board design process. These typically convert a circuit diagram (often called a schematic diagram) into a printed circuit board layout consisting of pads and tracks. The circuit board can then be drilled in the places where the various components are going to be located, for example, pad positions. These are the positions where the components will soldered to the copper track. Any unwanted copper (that is, copper not masked by the tracks, pads or other features in the printed circuit design) is removed. It is removed using either a photo-etch process, or for relatively small batches, a CNC milling machine. In the photo-etch process, the design is printed onto a transparent or semi-transparent sheet. This sheet is then carefully positioned against the ultraviolet (UV) sensitive layer of the circuit board material, used for the etching process. When UV light is shone onto the circuit board, some light is blocked by the pads, tracks and other features. Some light, however, is able to reach the copper not 'masked' by the circuit board design features. At this stage, the board is developed in a chemical to remove the resist material not 'masked' by the circuit design. The board is then fully immersed in a bubble etch tank which contains a corrosive etching chemical called ferric chloride. The board is left in the etch tank for about ten minutes or until the unwanted copper has been etched away, leaving just the pads and tracks.

Printed Circuit Board (PCB) design

Measurements relating to circuit board design tend to be in inches, where 0·1 inch = 2·54 mm, but some companies now use metric measurements.

Technical Terms

Track	thin layer of copper conductor that will conduct electricity
Pad	connection point for components; can be square, round or oval
Component side	the side of the board through which components are inserted
Solder side	the side with pads and tracks where soldering takes place
Rail	Main tracks: one is connected to the supply voltage (for example, the positive connection of a 12 V battery), and the other is connected to the 0V rail of the circuit (for example, the 0V connection of a battery or power supply).
Legend	a written information screen printed onto the component side
Schematic circuit	the circuit that is to be produced in symbol and connection format

Procedure for designing a PCB

It is good practice when designing a PCB to start by producing a circuit diagram (schematic diagram) of the circuit you want to manufacture. A systems diagram approach, where parts of the circuit are designed using block diagrams (see section 4·1), is a useful way to begin the overall design process. This helps to ensure that the main parts of the circuit are correctly integrated before you move to the circuit board design stage.

Printed circuit diagrams can be designed by hand using an etch resist pen to draw the circuit onto the board. You can also use transfers for pads, tracks and other features which can be transferred to the surface of the board. These produce an outline of the circuit required. There are also a range of computer software programs available for printed circuit board design work.

It is also good practice to make and test a circuit on a breadboard before making the printed circuit board, in order to make sure it works properly. If you are using a computer-simulated software program for the circuit board design, then the circuit can usually be tested using the software package.

Using simulation software has a number of advantages compared with making printed circuit boards by hand. Two advantages are that changes can be made quickly, and there is no need to start the process again if you make a mistake. It is easy to make mistakes when designing boards by hand. Using simulation software can therefore save you time. It can also allow for designs to be changed quickly and can help reduce material waste.

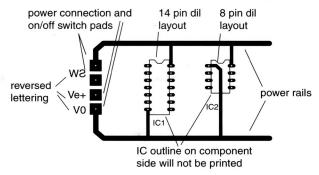

At the end of the process, the design should be checked to make sure it follows the schematic circuit precisely.

Component sizes are important:

Resistor	–	0·4 inch
Diode	–	0·3 inch
Preset	–	0·4 inch × 0·2 inch
Terminal	–	0·2 inch

0·1 inch is approximately 2·5 mm

Integrated circuits have pin connections with specific pin distances (spacings) between them. Pin spacings for an integrated circuit type can found in the data sheet for the integrated circuit you are dealing with. Most integrated circuits (ICs) have a 0·1 inch pin spacing between each in-line pin and a distance of 0·3 inches between the two rows of pins making the lines. For example, a 555 timer IC has eight pins and will require eight pads. The pads will be organised in two columns where each column has four pads. The columns are 0·3 inches apart. Because the two columns are in line with each other, the arrangement is called an 8-pin dual in line (dil) layout. Check component sizes accurately so that pad spacing will be correct.

△ Measuring pin spacing using digital calipers

Once the board has been made, it should be tested to make sure it has been manufactured correctly. It is also a good idea to add test points so that each stage of the circuit can be individually tested. These test points can be simple connection points for a multimeter or they can be small LEDs that light up if the circuit is working.

You should think about track widths and pad sizes during the printed circuit board design stage in order to make sure they can carry the current which will flow through them when used. A wider track or larger pad will be able to carry a higher current than a thinner track or smaller pad. Also, larger track and pad sizes give you larger surface areas for soldering the component.

The table below shows the safe maximum current for a range of track widths. For example, if a copper track has been designed to take a maximum current of 0.7 then the track width should be equal to (or wider) than 0.02 inches. This will make sure that the track can carry the current without overheating and becoming damaged in the process.

Track width (inches)	0.010	0.015	0.020	0.025	0.050	0.10	0.150
Maximum current (amps)	0.3	0.4	0.7	1.0	2.0	4.0	6.0

To help identify parts of a circuit, it is useful to include some written information on the board. This may be your initials, the name of the circuit or letters indicating where components should be go. This lettering is printed onto clear film or tracing paper. Remember that the writing on the film or tracing paper will need to be reversed (in mirror writing). If the print is held against the back of the circuit board, the writing will appear on the track side the right way round. Tracks should run horizontally or vertically as far as possible. Also, corners should be rounded so that sharp edges will not damage the wires and other parts when the circuit is assembled. Finally, check the finished design against the schematic circuit.

STRIP BOARDS

Strip boards are made by bonding copper tracks onto an insulated board. Each strip is perforated with holes for components to be mounted on the non-copper side of the board. The copper tracks can be broken at various points to produce a circuit. Components can be mounted either vertically or horizontally.

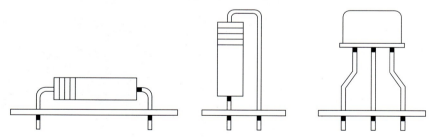

Finding faults on a stripboard circuit can be difficult and time consuming. Therefore, it is a good idea to use printed circuits for your projects, rather than strip board.

stripboard circuit - top view

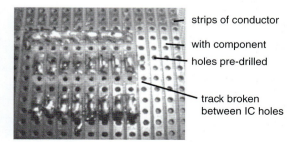

strips of conductor with component holes pre-drilled

track broken between IC holes

PROTOTYPE BOARDS

A traditional **prototype board** is often called a 'breadboard'. It can be used to experiment with different circuits during the early stages of design. During such design stages, faults can be 'ironed out' and different circuit configurations can be tried out. Components, wires and integrated circuit components are pushed into holes on the top of the board. They are held in place by solderless clips. This allows the components to be easily removed so that circuits can be changed quickly.

△ A prototype board

SOLDERING

Soft solder melts at a relatively low temperature (180 – 220°C). For electronic work, solder has flux running through its core. Flux is necessary to protect the metal from oxidising when the electrical parts are being soldered together. The flux helps to produce a reliable joint. Soft solder has contained lead until recently, but new legislation in many countries has meant that new soft solder with no lead content are now used in industry. This is because lead is toxic and continued use can be dangerous for workers.

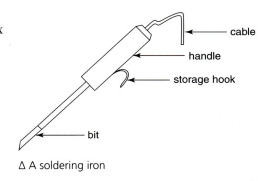
△ A soldering iron

Soldering technique

The technique of soldering is easy to learn if you remember these three basic rules:

- The component and connection both have to be heated beyond the melting point of the solder.
- The parts being joined have to be close to each other, because the solder will not bridge relatively large gaps.
- The soldering iron and parts to be soldered need to be clean. Oil, grease or oxidation on the wires and components to be soldered will stop the solder from flowing properly, during the soldering operation. Even sweat from your hands can have an effect, so make sure your hands are clean.

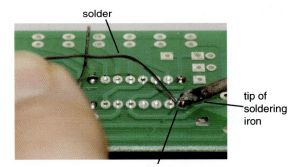
△ Soldering an IC

The soldering iron is used to heat the parts that are to be joined. When the soldering iron tip is hot enough to melt the solder, the solder is introduced to the soldering iron and joint from the opposite side of the soldering iron tip. Unwanted solder can be removed using a de-soldering tool sometimes known as a solder sucker. This uses a vacuum to suck up the unwanted molten solder. Braided copper wire can also be used to draw the unwanted solder away.

Health and safety

You need to be very careful when soldering, as it can be dangerous. Make sure you follow these guidelines before you begin, during the soldering process and after you have finished.

Before you begin:

- check the soldering iron cable for damage
- check the plug for damage
- check the tip of the soldering iron for any burnt areas
- use an extractor fan to take fumes away.

During the process:

- keep hands clear of the hot parts of the soldering iron. The tip temperature is approximately 200°C, which is twice as hot as boiling water so take care
- fumes from flux should not be breathed in
- any solder with lead content will pass on some of the lead to your hands, so make sure you wash them thoroughly after soldering
- the soldering iron should always be placed in the stand when it is not being used. Take care that the hot tip does not rest against the soldering iron cable by mistake.

After finishing:

- switch the iron off
- make sure that it is placed in the stand and remember that it will remain hot for some time. Remember that any components that you have been soldering will also stay hot for some time. The larger the soldered joint, the longer it will remain hot.

If you do burn yourself, wash the burn under cold water immediately and keep it under the water for a few minutes.

KNOWLEDGE CHECK

1. Briefly explain why printed circuit boards are commonly used for electronic circuits. (2)
2. Briefly explain why prototype boards are used for Design & Technology project work. (2)
3. What is the melting temperature range of electrical solder? (1)
4. Why has the alloy constituents of electrical solder been changed in recent years? (2)

KEY TERMS

electrical solder: an alloy with a low melting point, used to join electrical connections

printed circuit board (PCBs): a board with components mounted on etched tracks to form an electrical or electronic circuit

prototype board: a thin plastic board used to hold electrical components temporarily during the development of a circuit

strip board: a type of board used in electronic circuit construction which has a regular grid of holes positioned along parallel strips of copper

4·4·3 Switches

LEARNING OBJECTIVES
By the end of this unit you should:
- ✓ be able to recognise and be able to use various types of switch, including the micro switch
- ✓ understand the terms normally open (NO) and normally closed (NC) in relation to switching
- ✓ understand the uses of rotary and reed switches
- ✓ understand what is meant by the terms SPST and SPDT switches.

A **switch** is a device used to start or stop the flow of electricity in a circuit. Two common types of switch are toggle switches and push button switches. These are simple on-off switches which are usually operated by hand.

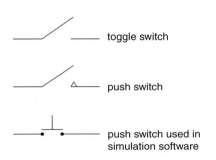

△ A toggle switch

△ The power buttons often found on remotes are push button switches.

PTM AND PTB STATES

The terms PTM (press to make) and PTB (press to break) relate to switching positions of press switches. For PTM, when the switch is pressed, the contacts of the switch are closed and the switch is switched on. A press to break switch (PTB) works in the opposite way. When the switch is pressed the contacts will open and break the circuit (so the switch is switched off).

△ A PTM switch

MICRO-SWITCHES

Micro-switches are a type of non-manually operated switch which can be switched on and off with a small force. Micro-switches are commonly used for applications such as safety switches for machine guards and industrial robot applications. They can also be controlled directly by cams, solenoids and other mechanism parts.

Micro-switches are very reliable and can operate numerous times without breaking.

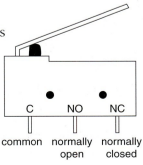

ROTARY SWITCHES   EXTENDED

Rotary switches can be used to switch a range of electrical circuits using one switch, for example, the selector dial on a digital multimeter. Switches can be 3 way, 4 way, 6 way and up to 12 way. These define the number of individual circuits the rotary switch can operate. A 3-way switch can therefore be used for three switching applications. There is always one common connection on the switch. For example, a 12 way switch has 13 contacts: the 12 ways plus the common connection.

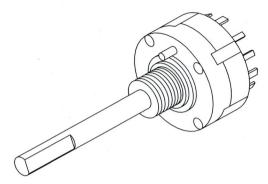

△ A 12 way rotary switch

REED SWITCHES

Reed switches use small magnets to operate them. Typical uses are revolution counters for shafts and spindles.

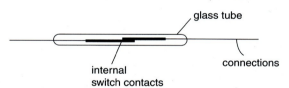

NORMALLY OPEN (NO) AND NORMALLY CLOSED (NC) SWITCHES

When a switch is **normally open (NO)** it means that the circuit will not be complete until the switch is operated (closed). A **normally closed (NC)** switch works in the opposite way. For a normally closed switch, the switch is on until switched off.

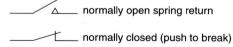

Some switches can be used either as a normally on or off switch depending on how they are connected to a circuit. For example, a micro-switch can have three connections: a common connection (which always has to be connected to a circuit); a usually closed connection; and a usually open connection. The switch can therefore be connected either way depending on how it is to be used.

The poles of a switch are the number of separate circuits it can operate. The throw describes how many positions each of the switches poles can be connected to. Common types are:

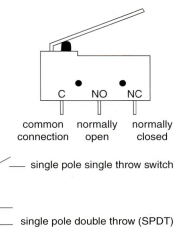

- **Single Pole Single Throw switches (SPST)**
A single pole single throw switch is the simplest type. The switch has just one input and one output. It can be used to operate one circuit and is either on or off.
- **Single Pole Double Throw switch (SPDT)**
A single pole double throw switch has three connections. One is a common connection. The others are normally open and normally closed. The switch can be used to either connect or disconnect the common terminal to another point in the circuit depending upon its switching position.

RELAYS

A **relay** is a special type of switch turned on and off by an electromagnet. When the electromagnet coil is activated, the relay contacts are brought together, connecting the common and normally open contacts. When the coil is deactivated, the common connects to the normally closed terminal. Relays are useful because they can operate a higher voltage or current circuit with a much lower voltage or current circuit. As such, they are often used to control mains circuits using a low-voltage battery operated circuit, often either 6 V or 12 V supply.

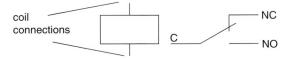

SPDT RELAY

A relay can also be used for special switching circuits such as reversing a motor and latching.

A latching action is when a brief signal or pulse from an input is held in place by a latch. This will happen in an alarm circuit where a sensor may be briefly activated, but the alarm output will stay on until the alarm is reset.

The circuits for both of these special uses are shown on the next page. Each circuit uses a double throw double pole relay (DPDT) relay.

△ Types of relays

Motor reverse circuit using a relay

A motor can be reversed by changing the wires powering the motor from one connection to the other. For example, if one wire is connected to the positive terminal and the other to the negative terminal, the motor direction can be reversed by changing round the original positive and negative connections.

A DPDT relay can be used to perform this activity automatically. A motor reverse circuit enables a motor to rotate in reverse, that is, anticlockwise. With SW1 open, the motor positive supply is fed to the top NC and bottom NO connections and the motor rotates clockwise, as shown in the first circuit. When SW1 is closed, the motor positive supply is fed to the top NO and bottom NC connections. This has the effect of reversing the direction of the motor, which now rotates anticlockwise.

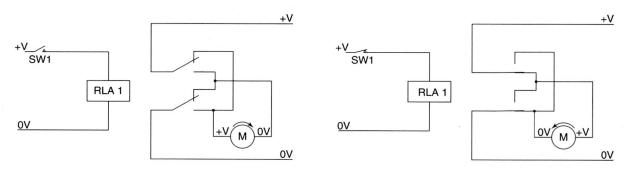

LATCH CIRCUIT

A latching circuit (latch) is a switch that once switched on will remain on until the power going to it is switched off. Latches are commonly used for alarm systems because once an alarm has been triggered, it often needs to be left on until the alarm has been attended to. A relay can be used to provide a latching circuit.

In view (a), SW1 is open and SW2 (the push to break switch) is closed. This will mean that under these circumstances the alarm bell is silent. However, when SW1 is switched on, it activates the relay coil as shown in diagram (b). The relay contacts close, causing the top contacts to switch the alarm bell on. At the same time, the bottom contacts will connect the supply to the coil. At this point, the coil is connected to the +V supply through SW1 and the coil contacts.

If SW1 is opened, the supply voltage to the coil will allow the alarm bell to remain on, regardless of the position of SW1. This is shown in (c). This is the latching effect.

To switch the alarm off, the supply to the relay coil must be interrupted by operating SW2, as shown in (d). At this point, the relay coil has no supply and the relay contacts are switched off.

Relay circuits of this type were used extensively in the past, but they now have fewer uses because they are more expensive than integrated or programmable integrated circuits.

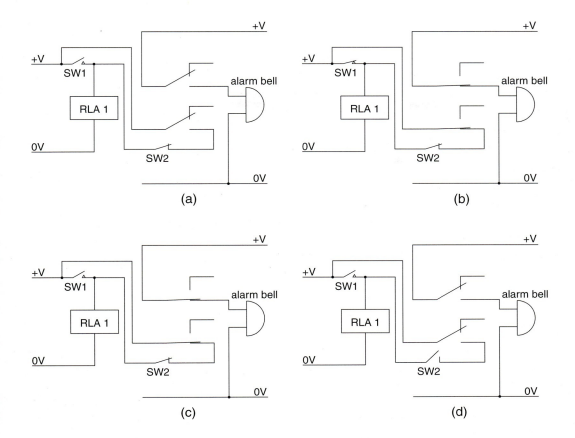

KNOWLEDGE CHECK

1. Give a use of (i) a toggle switch and (ii) a push switch. (2)
2. Give two applications of a micro switch in systems and control applications. (2)
3. Explain the terms *normally open* and *normally closed* in relation to switching. (3)

KEY TERMS

micro-switch: a highly sensitive small switch used in many systems and control applications (often operated by a lever or roller)

normally closed (NC): a switch state where the switch is on until it is operated

normally open (NO): a switch state where the switch is off until it is operated

relay: a special kind of switch often used to control a high current circuit using a low current input

switch: a device used to make or break an electronic or electrical circuit

4·4·4 Resistors

LEARNING OBJECTIVES

By the end of this unit you should:
- ✓ understand that resistors are divided into three main types: fixed resistors; variable resistors; and special resistors (thermistors, light dependent resistor).

FIXED RESISTORS

A fixed **resistor** limits the flow of electrical current flowing through it. During their operation resistors get hot. Resistors designed to handle large currents are physically larger because they need a larger surface area to dissipate the heat.

The unit of resistance is the ohm (Ω). Larger values are measured in kilohms (kΩ) = 1000 Ωs and Megaohms (MΩ) = 1000 000 Ωs. The symbol for a fixed resistor is:

△ Resistors

Colour coding of fixed resistors

Fixed resistors are identified by coloured bands on their surface.

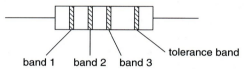

A colour code is used to identify resistors. The code to identify each band can be found from the following table:

Colour	First band	Second band	Third band (multiplier)	Tolerance
Black	0	0	× 1	red = 2%
Brown	1	1	× 10	gold = 5%
Red	2	2	× 100	silver = 10%
Orange	3	3	× 1000	
Yellow	4	4	× 1000 0	
Green	5	5	× 1000 00	
Blue	6	6	× 1000 000	
Violet	7	7	× 1000 000 0	
Grey	8	8	× 1000 000 00	
White	9	9	× 1000 000 000	

Example

A resistor that has the following coloured bands will have a **nominal resistance** of 5·6 kΩ and a tolerance of 10%.

TOLERANCE

Manufacturers cannot guarantee the exact resistance of resistors so, a **tolerance** value is given. Tolerance is the percentage error of a resistor and can either be 2%, 5% or 10%. For example, a 100 Ω with a 10% tolerance will have a value between 90 Ω and 110 Ω.

RESISTORS IN SERIES AND PARALLEL

Resistors can be connected in series or parallel. When doing so their combined **resistance values** can be calculated as follows:

Resistors in series

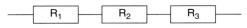

Total resistance $(R_T) = R_1 + R_2 + R_3$

For example, if resistors with resistance values of 100 Ω, 120 Ω and 220 Ω are connected in series, then the total resistance will be:

R_T = 100 Ω + 120 Ω + 220 Ω = 440 Ω

Fixed resistors are available in set values. The E12 series of values is usually used in schools and this consists of the following number values:

10   12   15   18   22   27   33   39   47   56   68   82

These values can then be multiplied by 100, 1000 or 10000 to give the values that are commonly used in circuits. For example, the E12 values × 100 gives 1kΩ, 1.2 kΩ, 1.5 kΩ, 1.8 kΩ etc. When multiplied by 1000 it gives 10 kΩ, 12 kΩ, 15 kΩ, 18 kΩ etc. The nearest E12 resistor to a set value never differs from that value by more than 10%. Small errors like this will rarely make any practical difference to the way that a circuit works.

SKILLS ACTIVITY

Determine the combined (total resistance) for the following resistors in series:

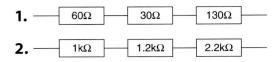

Resistors in parallel

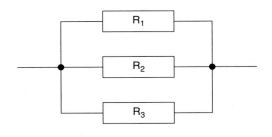

$$\frac{1}{R_T} = \frac{1}{R1} + \frac{1}{R2} + \frac{1}{R3}$$

For two resistors in parallel this can be simplified to:

$$R_T = \frac{R1 \times R2}{R1 + R2}$$

Example

Determine the combined resistance of the following network.

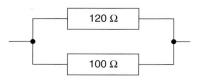

$$R_T = \frac{120\,\Omega \times 100\,\Omega}{120\,\Omega + 100\,\Omega}$$

$$R_T = \frac{12000\,\Omega}{220\,\Omega}$$

$$= 54.5\,\Omega$$

Other components, such as lamps and LEDs, can be connected either in parallel or in series. The advantage of connecting in parallel is that if one component fails, the remaining components will still operate. This can be particularly important when dealing with lighting: having all of the lights in a circuit fail could be dangerous.

In the case of LEDs in a parallel circuit, each LED will need a current limiting resistor. In a series circuit, it is possible to use a single current limiting resistor. One advantage of the series circuit is that the same current flows through each LED, meaning equal brightness from each LED.

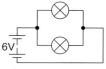
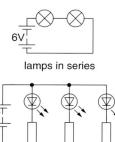
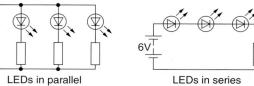

Potential divider

This arrangement of resistors divides the supply voltage (Vsupply) into two smaller values. The total value of the voltages across the two resistors (R_1 and R_2) will always add up to the supply voltage.

The arrangements shown below are commonly used, particularly in sensing circuits.

There are a number of ways of calculating the output voltage (Vout).

One of the easiest formulas to use is $Vout = \dfrac{R2}{R1+R2} \times Vin$

If R1 = 10kΩ and R2 = 18kΩ and the supply voltage is +9V

$$Vout = \frac{18}{10+18} \times 9 = \frac{18}{28} \times 9 = 5.79V$$

In the second circuit, a variable resistor and light dependent resistor (LDR) form the potential divider. If the light level gives a resistance of 2kΩ and the variable resistor (**VR1**) is set to 4kΩ and the supply voltage is +9V, the voltage between the centre point and 0V will be

$$Vout = \frac{2}{4+2} \times 9 = \frac{2}{6} \times 9 = \mathbf{3V}$$

In this circuit, VR1 is used to change the voltage caused by light falling on the LDR. It can therefore be used to alter the sensitivity of the circuit. The output voltage can then be used at a later stage in the circuit.

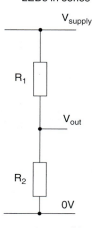
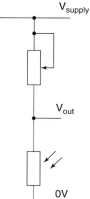

SKILLS ACTIVITY

Determine the total resistance of the network shown:

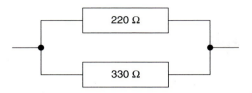

Variable resistors

Variable resistors, or potentiometers, have three principle uses:

1. To provide continual resistance change as the case of volume control.
2. To provide a resistance which can be adjusted to a particular value.
3. To provide voltage control by means of a potential divider circuit.

A variable resistor has three connections and a spindle to operate it. The spindle is turned to alter the component's resistance and can be cut to the length required so that knobs and other fittings can be added. Variable resistors tend to be secured to casings or similar features, and are connected to the circuit board using flexible wires.

△ A variable resistor

Variable resistors also come in pre-set resistor forms. The resistance value of a pre-set resistor is adjusted or 'set' using a small screwdriver when the circuit is being set up. Once adjusted, the pre-set is not normally adjusted again once the circuit in operation. Pre-set resistors tend to be soldered directly onto the circuit board. Both types of resistor are available with single-turn adjustment or with multi-turn adjustment. The multi-turn version is far more precise.

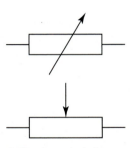

△ Circuit symbols for variable resistors

SKILLS ACTIVITY

What colour code would identify the following resistor values: (i) 220 Ω (ii) 1.2 kΩ (iii) 3600 Ω and (iv) 1.2 MΩ?

KNOWLEDGE CHECK

1. Give two uses of variable resistors in systems and control applications. (2)
2. Define the terms *nominal resistance* and *tolerance* when considering fixed resistors. (2)
3. If two resistors with resistance values of 330 Ω and 1 kΩ are connected in series, what will be their combined resistance? (2)
4. If the two resistors (330 Ω & 1 kΩ) are now connected in parallel, what will be their combined resistance? (2)

 KEY TERMS

nominal resistance value: a measure of value that states the resistance of a particular resistor

resistor: an electrical component designed to introduce a known value of resistance into a circuit

resistor tolerance: a measure of the permitted maximum and minimum resistance values of a particular resistor

4·4·5 Transistors

LEARNING OBJECTIVES
By the end of this unit you should:
- ✓ be able to recognise the collector, base and emitter connections of a transistor
- ✓ understand that transistors can be used as a fast operating electronic switch and as an amplifier
- ✓ understand how transistors can be used in simple sensing and timing circuits.

WHAT IS A TRANSISTOR?

A **transistor** has two main uses, to work as:
- a switch
- an amplifier.

A transistor has three connections. These are called the **emitter**, **collector** and **base**. It is important to be able to identify each lead because a transistor will be damaged if it is connected to a circuit incorrectly.

To help identify the leads, some transistors have a dot near the collector whilst others have a small tab to identify the emitter.

△ This type of transistor is often used for school projects.

HOW A TRANSISTOR OPERATES

The transistor has two circuits: an input circuit and an output circuit. Very often the emitter, or base circuit, is the input and the emitter, or collector circuit, is the output. This is known as a common emitter circuit.

The transistor is controlled by the base current. If there is no current flowing through the base/emitter circuit, no current can flow through the collector/emitter circuit. A small current can therefore be used to control a larger collector and emitter current. The gain of the transistor will determine the collector/emitter current for a given base/emitter current. In order for the current to flow there must be at least 0·6 V between the base and the emitter.

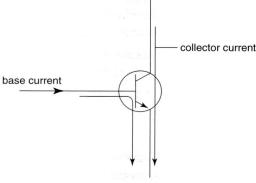

△ Transistor current flow

NPN TRANSISTORS

An **NPN transistor** is one of two transistor families. The other is a PNP transistor. NPN stands for negative-positive-negative and refers to the way the transistor is made in its silicon form. PNP stands for positive-negative-positive and has a slightly different internal silicon arrangement to make it operate. The correct name for a NPN device is bipolar junction transistor.

As with all transistors, an NPN type has three legs: base, emitter and collector. As it is a type of semiconductor component, it can be made to control the amount of electrical current flowing through it. NPN transistors often come in a discrete (single) package form. These are cased in either plastic or metal. They can also be part of an integrated circuit encasing a number of transistors. Modern integrated circuits in devices such as mobile phones hold tens of thousands of small transistors.

The NPN transistor is commonly used in amplification. It will amplify a small current to allow a larger current to flow through the output. The amount of amplification is known as the gain of the transistor. This is sometimes stated as hFE. When a small current is passed through the base-emitter circuit, a larger current will pass through the collector-emitter circuit. For example, a 1mA current passing through the base-emitter circuit will allow 50mA in the collector-emitter circuit, if the transistor has a gain of 50. As with many transistors, the base-emitter circuit requires about +0·6V at the base connection.

It is good practice to connect a resistor in series with the base connection to reduce the current flowing through the base, and consequently through the transistor itself. The resistor protects the transistor from too high a current, which might damage the transistor when operating.

The position of the legs can differ on a transistor. The best way to identify the leg positions is to refer to a manufacturer's catalogue which shows the pin diagram for a particular transistor. The leg positions will be typically marked as 'e', 'b' and 'c' to help identify the emitter (e), base (b) and collector (c) connections of the transistor.

An NPN transistor can also be used as a switch. As an electronic switch, it can operate many times faster than a mechanical switch and has no contacts to wear out.

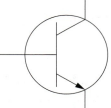

△ NPN transistor symbol

TRANSISTOR SWITCHES

Transistors are often used in switching circuits. For example, they can be used as the main switch component in a simple moisture detector which has a set of probes to detect the moisture, a light emitting diode (LED) to indicate if moisture is present and a transistor switch. When moisture is detected by the probes it will complete the circuit between each probe. The subsequent current flow will then operate the transistor by activating the transistor's base connection. This will allow current to flow through the collector and emitter and turn on the LED, as shown in the circuit opposite.

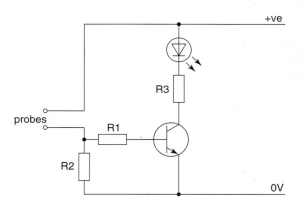

KNOWLEDGE CHECK

1. What are the names of the three connections on a transistor? (2)
2. What are the two main uses of a transistor? (1)
3. Draw a transistor circuit which could be used as a moisture detector. (3)

> **KEY TERMS**
>
> **base:** one of three pins of a transistor, used to control the switching or amplification action of the transistor
>
> **collector:** one of three pins of a transistor, normally connected to a positive part of a circuit
>
> **emitter:** one of three pins of a transistor, normally connected to a negative part of a circuit
>
> **NPN transistor:** An NPN transistor is one of two transistor families (the other being a PNP transistor). NPN stands for negative-positive-negative and refers to the way the transistor is made in its silicon form.
>
> **transistor:** a small three-contact electrical component used for fast switching or amplification

4·4·6 Diodes

LEARNING OBJECTIVES
By the end of this unit you should:
✓ be able to recognise the circuit symbol for a diode
✓ understand that a diode allows current to flow in only one direction
✓ understand how diodes are used to rectify current
✓ be able to describe the use of light emitting diodes.

WHAT IS A DIODE?

A **diode** is a semiconductor device that allows current to flow in one direction only. The circuit symbol for a diode is:

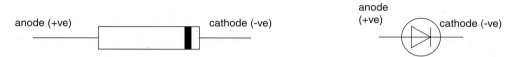

The direction that the current is allowed to pass is called the **forward bias**. The direction that the current is not allowed to pass is called the **reverse bias**.

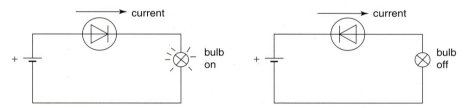

The diode has two leads called anode (positive connection) and cathode (negative connection).

USING DIODES TO PREVENT DAMAGE TO A CIRCUIT

A common use of diodes is to prevent damage when an output transducer with a coil is switched off. For example, either a relay or a buzzer contains a coil. When the power to the coil is switched off, a reverse current (back emf) can be generated, which can destroy a transistor.

A diode in reverse bias, with the cathode connected to the positive supply rather than the negative will act as a block to prevent this happening. An example of how a diode should be used is shown opposite.

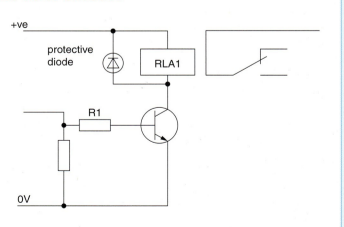

LIGHT EMITTING DIODES

A **light emitting diode (LED)** is a special kind of diode that gives out light. The LED is made from a semi-conducting material and it is possible to emit a range of colours including red, yellow and green light. LED arrays have now replaced light bulbs in a number of applications such as home and street lighting. Like all diodes, the LED will only allow current to pass through it in one direction. The amount of current is low and LEDs generally have to be protected using a series resistor so that they will not be damaged when used. The maximum amount of current for standard LEDs is approximately 10 mA. The precise maximum current can be found by referring to data sheets or catalogues.

△ A red light emitting diode (LED)

LEDs can differ in size, colour, shape and intensity. The type of light output can be visible, infra red or ultra violet. Infra red LEDs are commonly used in remote control devices.

When calculating the value of the fixed resistor used to protect an LED, the voltage drop across the LED must be taken into account. This is normally given in the manufacturers' data sheets, but as a guide, it is approximately 2V. An arrangement of Ohm's Law can be used for the calculation i.e. $R = V \div I$.

△ Electronic circuit symbol for an LED

For example, if the supply voltage is +9V, and the voltage drop across the LED and current flowing through the LED are 2V and 10mA respectively, then

$$R = \frac{9-2}{0.010} = \frac{7}{0.010} = 700$$ The nearest resistance value to this is 680Ω.

SEVEN SEGMENT DISPLAYS

Seven segment displays are a common form of display used for a wide range of applications such as digital clocks, digital radios, car dashboard displays and similar display features.

A seven segment display is made up of seven separate LEDs: one for each segment and, in most cases, an eighth LED is used to show a decimal point. The display shown below is a typical example of a single digit display. The pin diagram shows what each pin is used for.

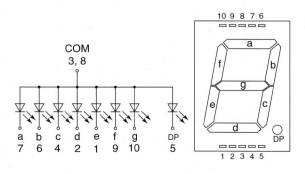

Pin No.	Function
1	segment e
2	segment d
3	common
4	segment c
5	DP
6	segment b
7	segment a
8	common
9	segment f
10	segment g

△ A 7 segment display component

A display either has a common anode or a common cathode. This means that one leg of the LED is connected, either as an anode or cathode depending on what type of LED is used. The example shown is a common anode type. This is because all the anodes (the +ve side of each of the eight LEDs) are connected to the common connection identified as COM on the diagram. As you can see, 9 connecting wires will be needed - one for each of the seven number segments, one for the decimal point and one for the common connection.

The pins are laid out in a similar way to other ICs - dual in line (dil). As with any LED, a protective resistor is needed for each segment of the display. Separate resistors can be used to protect each segment. You can also use a resistor array which has either seven or eight resistors built into a dil package, as shown opposite. Using a resistor array of this type tends to be easier and neater to use because the resistors are neatly housed in an enclosed package and there is no need to attach individual resistors to the LED display connections.

△ A resistor array

KNOWLEDGE CHECK

1. What is the main purpose of a diode in a circuit? (1)
2. What are the leads of a diode called? Which one is the positive connection and which one is the negative connection? (2)

KEY TERMS

diode: an electronic component which allows current to flow predominantly in one direction only

forward bias: the main direction of current flow through a diode

light emitting diode (LED): a type of diode which emits light when operating

reverse bias: the direction where there is no current flow until the diode breaks down

4.4.7 Transducers

LEARNING OBJECTIVES
By the end of this unit you should:
- ✓ be able to describe the use of an LDR
- ✓ be able to describe the use of a thermistor
- ✓ understand that thermistors can have either negative or positive coefficients.

LIGHT DEPENDENT RESISTORS

A **light dependent resistor (LDR)** is a type of resistor which changes its resistance relative to the amount of light falling on it. An LDR is shown below with its circuit symbol.

As the light falling on the LDR increases then its resistance decreases. A typical LDR has a resistance of approximately 120 Ω when a large amount of light is falling on it, going up to 10 MΩ when it is completely covered.

△ A light dependent resistor (LDR)

Using light dependent resistors in circuits

Typical uses of LDRs are for operating light detector or darkness detector circuits.

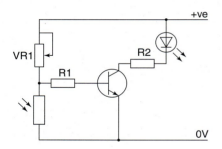
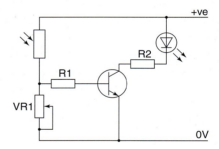

The light detector circuit on the left will switch the LED on as it gets dark. The level of darkness when it switches is controlled by VR1, which forms a potential divider circuit with the LDR. As the resistance in the LDR increases, the voltage going into the transistor base through R1 will increase. A point will be reached when the voltage into the base goes above the 0.6 V switching voltage of the transistor, which will switch the resistor on.

The circuit on the right works in the opposite way by switching the transistor on as the light level increases. The increased light level reduces the resistance at the top of the potential divide circuit and, as a consequence, the voltage going into R1 and the transistor base increases. In both cases a series R2 has been included in the circuit to reduce the amount of current flowing through the LED. This protects the LED from damage due to too high a current flowing through it.

THERMISTORS

A **thermistor** is a resistor which is designed to change its resistance with temperature change. Thermistors can have negative temperature coefficients or positive temperature coefficients. By far the most common is the negative coefficient or NTC thermistor. In the case of the NTC thermistor, the resistance will fall with a rise in temperature. For a positive coefficient, the resistance will rise with a rise in temperature. The symbol for an NTC thermistor is shown opposite. The −t° is sometimes missed off the symbol. When a thermistor is added to a circuit like the one opposite (replacing the LDR), the temperature can be controlled.

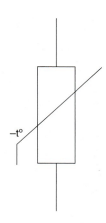

STRAIN GAUGE   EXTENDED

Another type of transducer is the strain gauge. These are often used as part of measuring equipment such as weighing scales, or for measuring the extension and deflection of structural parts such as building columns, rail tracks and aircraft wings.

A strain gauge is a small component which works as a resistor. It is generally made by sticking a thin foil of resistance material onto a suitable backing material. A gauge can then be stuck to the part it is to be used on, using a suitable adhesive. Strain gauges are physically small components and tend to have a nominal resistance of 120 Ω. In practice, strain gauges form part of a circuit called a Wheatstone bridge. In the simplest case (that is, using one strain gauge for one 'arm' of the bridge), the strain gauge takes the place of one resistor in the bridge and becomes part of a pair of potential dividers. If there is no extension, compression or bending of the strain gauge, the voltmeter connected to the central part of the bridge should register no difference in voltage potential between the two sides of the bridge. When tension, compression or bending is applied to the strain gauge, its resistance will change in relation to the change in dimensions. This will result in a change in voltage at the resistor junction, which will change the voltmeter reading accordingly.

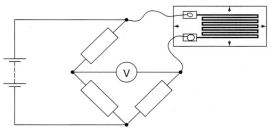

△ Wheatstone bridge circuit

KNOWLEDGE CHECK

1. What is meant by the term transducer? (1)
2. Explain the principle of a light-dependent resistor. (2)
3. Explain the principle of a thermistor. (2)

KEY TERMS

light dependent resistor (LDR): a type of resistor that changes its resistance in response to the amount of light falling on it

thermistor: a type of resistor in which the level of resistance depends on its temperature

4·4·8 Capacitors

> **LEARNING OBJECTIVES**
> By the end of this unit you should:
> ✓ understand the units of capacitance
> ✓ understand that capacitors are used for timing applications
> ✓ understand the difference between polarised and non-polarised capacitors.

WHAT IS A CAPACITOR?

A **capacitor** is a device which is able to store electricity. Capacitors are also used in timer circuits. The larger the capacitance value the more electricity it can store. The unit of capacitance is the **farad** (F), which is the standard unit of capacitance in the International System of Units (SI). In practice this is quite a large value for electronic component purposes, and smaller values tend to be used. Common units of capacitance for electronics are:

- microfarads (μF) = 1 ÷ 1 000 000 F
- nanofarads (nF) = 1 ÷ 1 000 000 000 F
- picofarads (pF) = 1 ÷ 1 000 000 000 000 F

TYPES OF CAPACITORS

Capacitors are divided into two groups: polarised and non-polarised. In each case the capacitor has two connections.

△ There are various kinds of capacitors and NIT resistors.

Polarised capacitors

Electrolytic capacitors and tantalum-bead capacitors are polarised. These generally have larger capacitance values than non-polarised capacitors. **Polarised capacitors** have a positive (anode) and negative (cathode) connection, and these have to be connected the correct way to circuits. Electrolytic capacitors are either axially or radially mounted. In many cases the polarity and capacitance values are shown on the capacitor, or, in the case of radial capacitors, one leg is shorter than the other.

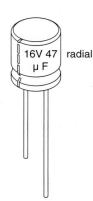

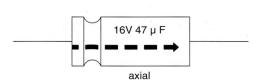

Non-polarised capacitors

Non-polarised capacitors are sometimes called non-electrolytic capacitors. They are usually smaller than the electrolytic type and are named according to the material of manufacture; for example, polystyrene, ceramic, polyester. The capacitance values range from a few picofarads to several thousand microfarads. Non-electrolytic capacitors can be connected either way round in a circuit.

Capacitors can be charged through a resistor to control the speed of charge. This is shown in the circuit below. The graph shows the voltage at point X as the capacitor charges after SW1 has been closed.

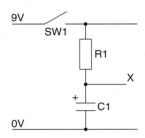 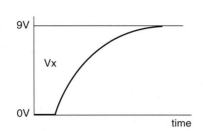

The effect of discharging a capacitor through a resistor can be seen in the next circuit. When SW1 is closed the capacitor can discharge through R1. The graph shows the value of VX as the capacitor discharges.

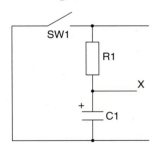 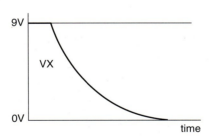

KNOWLEDGE CHECK

1. What is the SI unit of capacitance? (1)
2. Name the two basic types of capacitor. (2)
3. What do capacitors do in a circuit? (2)

> **KEY TERMS**
>
> **capacitor:** a two terminal component used to store an electric charge temporarily
>
> **farad:** the unit for measuring capacitance (the ability to store electricity)
>
> **non-polarised capacitor:** a small capacitor whose two connections can be connected either way round in a circuit
>
> **polarised capacitor:** a capacitor that has positive and negative connections and which must be connected the correct way in a circuit

4·4·9 Time delay circuits

LEARNING OBJECTIVES

By the end of this unit you should:
✓ be able to recognise a delay circuit using discrete components   **EXTENDED**
✓ be able to recognise the pin sequence of an integrated circuit
✓ be able to use a 555-timer for astable and monostable timing applications
✓ be able to calculate the time periods for astable and monostable operations.

DELAY CIRCUITS   **EXTENDED**

Time delay circuits are often made using **discrete components** such as transistors, resistors, capacitors and LEDs or lightbulbs. The example below shows a typical circuit drawn using a computer simulation package. This has the advantage that the circuit can be tested and modified relatively quickly using the computer software rather than making changes by hand.

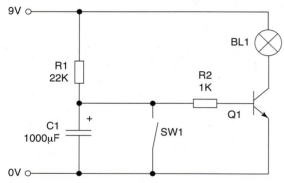 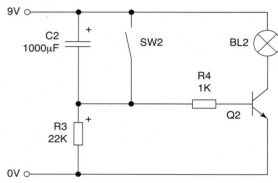

The circuit on the left has a lamp that is off until switch SW1 is opened. At that point, the capacitor begins to charge through resistor R1. When the capacitor is charging, the voltage across the capacitor will increase over time. When it reaches around 0·7V, the transistor base will switch on. When SW1 is closed, the capacitor will immediately discharge and the lamp will go out.

The circuit on the right works in the opposite way, and will cause a delay in the lamp switching off when SW2 is opened. The negative side of the capacitor is at 9V when the switch is opened. R3 will gradually cause the voltage in the capacitor and the voltage flowing into the transistor base to drop. Once the voltage has fallen below 0·7V, the transistor, and consequently the lamp, are switched off. In practice, this type of circuit is not used very often. Monostable integrated circuits or programmable ICs tend to be used because they are more accurate.

To determine the time it takes to charge a capacitor with a series resistor, the formula $t = CR$ can be used where t is in seconds, C in farads and R in ohms. In the examples above, a 22kΩ resistor is charging a 1000μF capacitor. The time taken to fully charge the capacitor is 0·001F × 22000 = 22 seconds. Because the transistor switches on well before the capacitor is fully charged, the actual delay will be far less than 22 seconds.

INTEGRATED CIRCUITS

Integrated circuits have developed extremely rapidly over the last few years and many of today's products have such circuits to help control their operation. An integrated circuit is a tiny piece of semi-conducting material, which has a complete electronic circuit etched onto it. An integrated circuit is often called a 'chip'. The most common material used to make these circuits is silicon and are therefore named silicon chips.

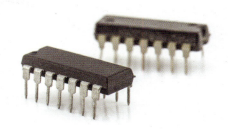

△ An integrated circuit is often called a 'chip'.

The advantages of using integrated circuits compared with circuits made up of individual components soldered on a circuit board include:

- they are extremely reliable
- they take up a small amount of space so products, for example mobile phones, can become increasingly smaller
- they are relatively cheap to produce
- they require less power to operate which helps reduce the product's operating costs.

555-TIMER INTEGRATED CIRCUIT

The 555-timer chip was one of the first integrated circuits used for timing and time delay applications. This integrated circuit can be used as either a monostable timer (one shot timer) or as an astable timer.

The 555-timer chip comes as an 8-pin dual-in-line package where pin number one is identified by a notch or a dot.

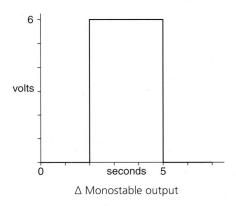

△ Monostable output

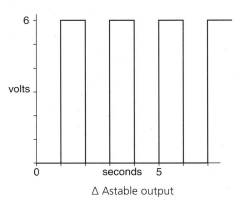

△ Astable output

When the monostable timer is switched on, the output pulse will suddenly go from its 0V state to 6V after a two second delay. This is the start of the **monostable output** state. After a certain time, the output pin (3) changes from 0V to the supply voltage (in this case +6V), and the output returns to its 0V state.

In the **astable output** graph, the pulses are continuously on and off i.e. on - wait 1s - off - wait 1s. In this case, the complete cycle takes 2 seconds. Pulse rates are normally given in Hertz (Hz). A rate of one cycle per second is 1 Hz. The pulse shown is half this speed so is 0.5 Hz. This is a very slow pulse rate and the output on the LED could be seen easily by eye. Much faster rates can be produced using a 555 IC: 1000 cycles per second = 1000 Hz 1kHz (kilo hertz) and 1000000 cycles per second = 1 MHz (Mega hertz).

The 555 as a monostable timer (one-shot timer)

Below is the circuit diagram for this type of timing operation. For any IC circuit, it is useful to have access to the manufacturer's data sheet. This will give full details of circuits that can be used with the device. A web search will normally give access to the datasheet. For example, a search for '555 timer datasheet' gives at least four versions from different manufacturers of the device.

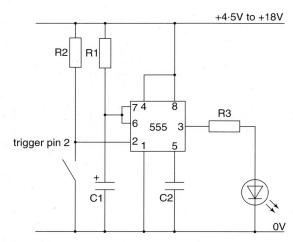

In this case, when the switch (S1) is pressed a square wave pulse will be output at pin 3. This will light the LED for a certain amount of time, and then the pulse will stop.

The time for one pulse (the one-shot) of the LED can be estimated from the relationship:

$T = 1.1 \times R1 \times C1$

Where R1 = the resistance value of the resistor labelled R1 in the circuit.

C1 = the capacitance value of the capacitor in the circuit.

Therefore, the time of the one-shot pulse can be changed by altering the values of R1 and/or C1.

The purpose of R2 is to protect the LED from blowing due to the pin-3 being too high.

SKILLS ACTIVITY

Here is a 555-timer. If resistor R1 is 1 MΩ and C1 is 100 μF, how long will the LED remain on?

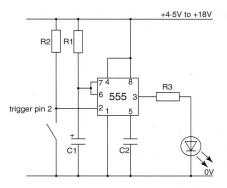

The 555 integrated circuit used as an astable oscillator

In a **555 integrated circuit**, the LED connected to pin 3 (the output connection) will flash on and off continuously until the timer is switched off.

The rate of flashing depends on the values of R1, R2 and C1. Changing the combination of these values will change the rate of flashing. As with the monostable circuit, the R3 resistor is there to protect the LED from being damaged when the timer is in use.

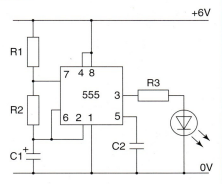

The time period for flashing can be determined from:

Time (seconds) = $0.7 \, (R1 + R2) \times C1$

Programmable ICs

Programmable integrated circuits are components which can be programmed to carry out a range of tasks. Programmable ICs programmed for a specific purpose are now found in most electronic products including cell phones, microwave ovens, washing machines, car control circuits, alarm systems and manufacturing control systems. Because programmable ICs have an integrated logic system and are flexible enough be programmed for such purposes, they have largely taken over from building circuits that use dedicated logic ICs, for example, circuits made from dedicated AND, NAND, OR and other gate types (logic is covered in more detail in section 4.4.10). As such, a programmable IC will typically use fewer components, and is easier to build and adjust than a dedicated logic circuit. They have the advantage that extra inputs and outputs are usually available and the cost of the ICs is low when compared to a dedicated circuit.

Advantages

- Lower parts count
- Easier adjustment to the circuit
- Easier fault finding
- Standard main section to the circuit
- Allows more complex problems to be tackled
- Program can be quickly changed.

Disadvantages

- Programming skill has to be developed
- Real circuit does not always do what is expected
- Precise supply voltage required
- Very often leaves surplus inputs and outputs that are not required.

PIC ICs (programmable interface controllers) are part of a family of microcontrollers that are quite easy to use. They vary from 8 pin controllers with up to six inputs and outputs, up to a 40 pin controller which has up to 33 inputs and outputs.

When used for time delays circuits, the PIC is extremely accurate and can count directly in milliseconds or seconds. The 8 pin PIC can hold over 1700 lines of program, more than enough for a school project. Supply voltage for many of these devices is less than 3V.

KNOWLEDGE CHECK

1. What is meant by a monostable output? (1)
2. What is meant by an astable output? (1)
3. What is the formula for determining the time for one pulse of a 555 timer used in its monostable mode? (3)

KEY TERMS

555 integrated circuit: a type of integrated circuit used for a variety of timing applications

astable output: an output that changes between being on and off while the circuit is operating

discrete components: separate individual components such as resistors, capacitor, transistors and diodes

integrated circuit: an extremely small circuit incorporating many components, etched onto a piece of semi-conducting material

monostable output: an output that is on for a specified length of time while the circuit is operating

PIC ICs (programmable interface controllers): an integrated circuit that can be programmed to carry out a task or range of tasks

4·4·10 Logic gates and operational amplifiers

LEARNING OBJECTIVES
By the end of this unit you should:
✓ understand the truth tables for NOT, AND, NAND, OR and NOR gates
✓ understand the difference between TTL and CMOS logic circuits
✓ be able to use CMOS logic circuits in project work
✓ recognise and use operational amplifier (Op Amp) integrated circuits in project work
✓ be able to calculate the gain of an inverting operational amplifier
✓ understand how to use an op-amp as a comparator.

LOGIC GATES

Logic gates are switching circuits where the state of their output depends on the state of their inputs. There are several types of logic gates; the most common are called NOT, AND, NAND, OR and NOR gates. XOR gates are not so commonly used and there are very fast acting gates called 'Schmitt' gates that are useful for sensing circuits.

Each gate has a truth table which states the logic gate's output for all the possible input states. In the truth table, a high state (or logic 1) refers to a voltage close to the supply voltage present at either an input or output. A low state (or logic 0) is where the voltage to or from the gate is at zero volts (0 V) or near to it (for example, 0·05 V).

NOT gate or inverter
A NOT gate is the simplest gate, and has only one input and one output. The gate produces a high as output when the input is low, and vice-versa.

Input	Output
0	1
1	0

NAND, AND, NOR and OR gates
All these type of gates can have two or more inputs, however, they are commonly shown in a truth table with two inputs. The truth table for the 2-input gate circuits are shown below.

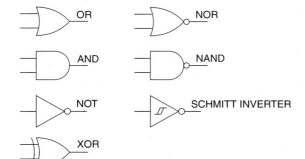

A	B	OR Q	NOR Q	AND Q	NAND Q	NOT Q	XOR Q
0	0	0	1	0	1	1	0
0	1	1	0	0	1		1
1	0	1	0	0	1		1
1	1	1	0	1	0	0	0

INTEGRATED CIRCUITS LOGIC GATE FAMILIES

Integrated circuit logic gates are available in different types and manufactured using different technologies. The most common types found in schools are TTL (Transistor Transistor Logic) and CMOS (Complimentary Metal Oxide Semiconductor) type.

TTL integrated circuits are sometimes known as 74 series, because their code number begins with 74. In a similar way CMOS integrated circuits are known as 4000 series. In practical terms, the TTL series requires a precise 5V supply. However, CMOS ICs can operate with a supply ranging from 3–15V.

The two types of IC cannot be mixed in a circuit. Each IC will have a pin diagram to show the connections. These will differ between the logic families. The diagrams below show the pin diagrams for a quad 2-input NAND gate in both versions. The majority of logic ICs used in school will have either 14 or 16 pins in a dual in line arrangement.

Pin 1 is always on the left hand side of the end with the notch or circle. Numbers follow down the left continuing anticlockwise up the right. Numbering for all ICs is done in this way.

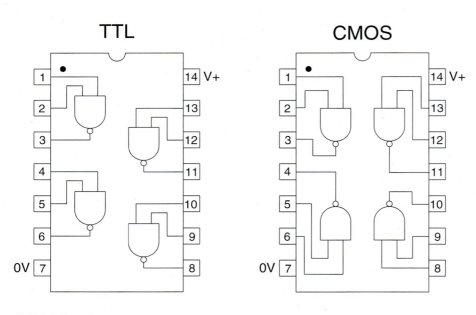
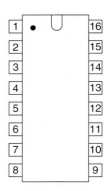

CMOS logic gates

Logic gate circuits are often used as the central part of alarms systems in school projects. The system shown on the next page has six inputs to OR gates. If any one of the six inputs are high, this will cause a high at the output.

To make this system, you would need two quad OR gate ICs. There are only five gates needed in the system so that would leave three unused gates. Spare gates can often be useful for another part of a circuit. With CMOS ICs any spare input gates should be connected to either the supply voltage or to 0V otherwise the circuit may not operate properly.

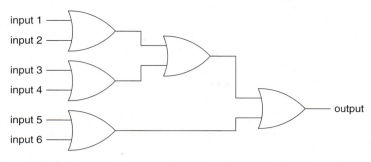

Another practical example using logic is shown below:

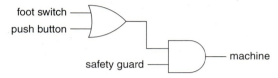

A machine that can be switched on using either a foot switch or a push button must have the safety guard in position before it will operate. This can be done using an OR gate with the output feeding into an AND gate as shown. When it is turned into a circuit using simulation software the result is shown below.

All three switches are connected normally low (0V) through a resistor. When the switch is operated the signal to the logic gate goes high (9V). This is because a logic gate does not recognise any signals other than high or low.

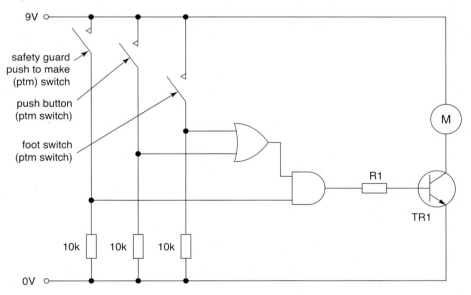

It is difficult to know what is happening in a logic circuit. Test equipment (such as a logic probe) can be used, but an easier way is to build test points into the circuit as shown opposite. When the output of the NAND gate is high the LED will be on.

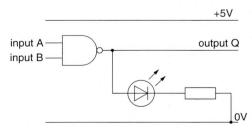

LOGIC CONTROL SYSTEMS IN EVERYDAY LIFE

Logic control systems are found in many contexts in our everyday lives. Below are some examples of logic gate circuits which can be used as the basis for simple practical projects. Logic gates are a useful way to achieve certain outputs because they allow logical procedures to be built into the circuit design. For example, in the greenhouse control example, the circuit allows for both the temperature and water levels to be monitored. If these are not at the required levels then an alarm will sound, alerting the greenhouse owner to the problem. Other everyday applications include heating control for houses, traffic light control systems and counting systems.

> **i)** The circuit below simultaneously controls a set temperature and a set moisture level for a greenhouse. The moisture sensor could simply be two wire probes which act as a switch if there is moisture present between them. The probes would typically be placed in the soil where the plants are growing. If the soil becomes too dry, then there would be no electrical flow through the probes. If a set moisture level was required, a moisture detecting switch could be used. The set temperature level is achieved using a thermistor and variable resistor as a potential divider network, so that an input signal to the NAND gate occurs at a particular temperature setting. If the temperature drops below this level, the signal to the AND gate will stop and the NAND gate will be switched on. In a similar way, when the moisture level becomes low, the NAND gate will also switch on, operating the alarm buzzer.

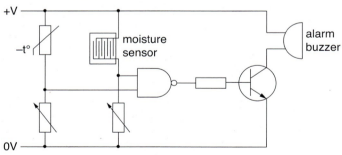

△ Controlling temperature and moisture levels in a greenhouse

> **ii)** This example shows a simple burglar alarm circuit which is used with two switches. One of the switches could be placed on a door frame and the other switch on a window. When one of the switches is broken by a burglar breaking either one of the switches, the LED will light up. The LED could be part of a control panel which monitors a range of burglar systems. Each switch is in its open-to-break mode, so that while they are both closed, the NAND gate will not operate. Once one of the switches is broken, the NAND gate will switch on, operating the LED.

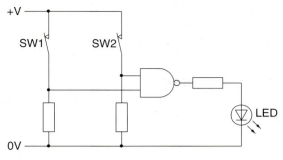

△ Burglar alarm circuit

iii) There are many case where you might like a doorbell which works during daylight hours only. This might be because you do not want anybody to call at your home or place of work once it is dark for personal safety reasons. This circuit makes use of a doorbell switch and a LDR that detects the lighting level of the surroundings. The lighting level is set by adjusting the LDR potential divider output (using the variable resistor), so that when it is light there no input to NAND gate 1. This means that when it gets dark, there will be an input to NAND gate 1. If the door bell is pressed when it is dark, there will be two input signals to the NAND gate. This means that the NAND gate output will be zero, and therefore there will be no output to the AND gate joined in series. In this mode, the transistor will also be switched off. However, if the doorbell is pressed during the day when it is light, then NAND gate 1 will be able to operate. In so doing, the AND gate inputs both become high and, as such, the AND gate output is able to switch on the transistor and bell part of the circuit.

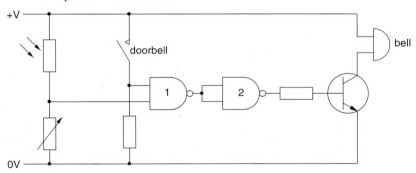

△ Doorbell which works in daylight

OPERATIONAL AMPLIFIERS (OP-AMPS)

Operational amplifiers or **op-amps** are used to amplify small input voltages, or they can be used to compare two voltages. The 741 op-amp is one of the most common types and is used in a wide range of amplifying applications, particularly in the instrumentation field. For example, they are used to amplify output voltages from

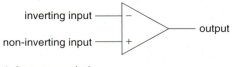

△ Op-amp symbol

strain gauge circuits, to amplify blood pressure measurements in blood pressure instruments, to amplify the temperature output of boiler heating systems and as part of the control systems for aircraft and other forms of transport. The symbol for the amplifier is:

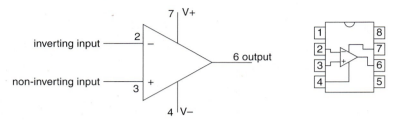

The 741 is a dual in-line package and details of the pin connections are shown below.

Pin number	Connection
1	Offset null (generally not used)
2	Inverting input
3	Non-inverting input
4	Negative (−ve) supply or 0v
5	Not generally connected (offset null)
6	Output
7	Positive (+ve) supply
8	No connection

The maximum supply voltage to the 741 op-amp is 16 volts.

VOLTAGE GAIN OF AN AMPLIFIER

In general, op-amplifiers are used to convert a small voltage change at the input to a larger voltage change at the output. The amount the amplifier amplifies the voltage is called the voltage gain.

$$\text{Voltage gain} = \frac{\text{change in output voltage}}{\text{change in input voltage}}$$

Example

If the change in input voltage is 10 millivolts and the change in output voltage is 1 volt, you would determine the voltage gain of the amplifier like this:

$$\text{Voltage gain} = \frac{\text{change in output voltage}}{\text{change in input voltage}}$$

Voltage gain = $\dfrac{1V}{10\ mV}$

Voltage gain = $\dfrac{1V}{0\cdot 01V} = 100$

INVERTING AMPLIFIER `EXTENDED`

A common way to use an op-amp is in its inverting mode. Here the output is an inversion of the signal entering the amplifier (that is, when the input voltage becomes more positive, the output becomes more negative). An advantage of using the amplifier in this way is that it increases the stability of the output circuit. The connections for an inverting amplifier is shown below.

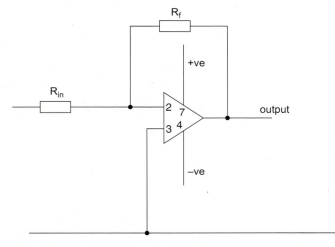

For this type of application, the gain depends on the ratio of the two resistors that are connect to the circuit: R_{in} (input resistor) and R_f (feedback resistor).

Amplifier gain = $\dfrac{-R_f}{R_{in}}$

Example

Calculate the gain when the input resistor = 1 k and the feedback resistor = 10 k.

Voltage gain = $\dfrac{-10\ k}{1 k} = -10$

This means that the output voltages will be ten times the input voltages, but the voltage polarity will be inverted.

SKILLS ACTIVITY

Fill in the missing values.

R_f	R_{in}	Gain
5600 Ω	1200 Ω	
2.2 kΩ		5
	1.2 kΩ	10

COMPARATORS

A 741 amplifier is often used as a **comparator**. A comparator is a circuit which compares two input voltages. When used as a comparator, the op-amp will compare the voltages between pins 2 and 3. If the voltage at the inverting input (pin 2) is greater than the voltage at the non-inverting input (pin3), the output will be close to 0V.

If the non-inverting input is greater than the inverting input the output will be close to the supply voltage. The change between the two voltages is very rapid and is useful for detecting voltage levels from sensors. An alternative IC that is a direct replacement for the 741 is a CA 3140 IC. This uses less power.

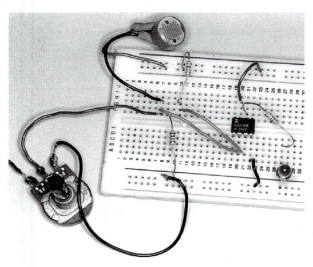

△ A comparator

An example of how an op-amp can be used with a light dependent resistor (LDR) to sense a set input level of light is shown below. For a circuit like this, a variable resistor (R1) is used to set the reference voltage for the lighting level required. This can be done using a voltmeter. An LED is connected to the output pin to indicate when the comparator circuit 'switches on', that is, when the +ve goes above the reference voltage. This type of circuit is also useful for temperature control circuits.

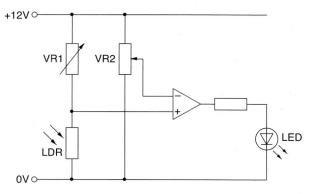

△ An op-amp can be used with a light dependent resistor (LDR) to sense a set input level of light.

KNOWLEDGE CHECK

Draw the truth tables for:
1. a **NOT** gate (2)
2. a **NOR** gate (2)
3. a **NAND** gate (2)

> **KEY TERMS**
>
> **CMOS logic gate family:** a category of logic gates (4000 series) commonly used for school projects
>
> **comparator:** a circuit which is designed to compare two input voltages and is often used for sensing purposes
>
> **logic gate:** an on/off switch that is the basic building block for electronic logic circuits
>
> **operational amplifier:** a type of integrated circuit used to amplify low voltage input signals

The coursework project is the part of the course that is often seen as the most enjoyable and exciting. Most people choose Design & Technology because they like designing and making things. The design project will give you the opportunity to use a wide range of graphic media, tools and equipment. The project has two elements: the portfolio and the product you make. It is important that you put evidence of your work into the design portfolio, so do not throw anything away.

You will also need to provide good quality, clear photographs of your design throughout the making process, and photographs of the final outcome, to demonstrate your skill and the quality of the item.

This unit will help you plan your work and you will see examples of student portfolio sheets. These are just guides, however, and it is important that you follow the instructions given to you by your teacher.

Work hard on this project and you will make a product to be proud of for a long time. A good project and portfolio can also be useful when talking to prospective employers or applying to universities, as it will demonstrate that you have a wide range of skills, knowledge and understanding. Good luck and enjoy the creative process!

STARTING POINTS

- You will focus on a theme from either Graphic Products, Resistant Materials or Systems and Control.
- You will present your project on A3 paper.
- The theme will be negotiated with your teacher and school.
- Try to incorporate CAD/CAM where at all possible.

SECTION CONTENTS

Design Project requirements
Stages of the design process:
1 Identifying a need and establishing a design brief
2 Researching the design brief and developing a specification
3 Generating and exploring design ideas
4 Developing a proposed solution
5 Planning for production
6 Product realisation
7 Testing and evaluation

5 The Project

The Project

GUIDANCE FOR COMPONENT 2: THE PROJECT

As part of your studies for the Cambridge IGCSE/O Level Design & Technology course, you will need to carry out a design-and-make activity. The project is an important part of the course, and perhaps the most enjoyable, as you have the opportunity to design and make a product that you feel best reflects your abilities and creativity. It is your opportunity to demonstrate everything you have learnt during the course.

Your project must be unique to you, and you should focus on one of the following options:

- Graphic Products
- Resistant Materials
- Systems and Control.

You are free to choose the theme for your project, and your teacher might also give you advice to help you decide. You will usually work on your project at school during the final two terms of the course, but this may differ from school to school.

Even though your project will be based on your chosen option (Graphic Products, Resistant Materials, or Systems and Control), you may want to use some of the knowledge and skills you have learnt from the other options as well. This is fine, but not a requirement. You must remember to stay focused on the option you have chosen, without too much overlap with other options. For example, you may be focusing on Systems and Control but, due to the nature of Design & Technology, you will need to house your circuitry in a casing made from resistant materials.

Your final project should include a design portfolio and practical outcome. You can use CAD/CAM if your school has these facilities. Even so, you should still present all of your relevant portfolio work in hard copy as an A3-size folder. Your teacher will explain that electronic submissions are not acceptable.

The portfolio must include photographs of your made product, including an overall view together with views showing all the finer detail.

If, for example, you chose a Graphic Products option, the folder should contain all the preliminary design work for your made product, which should be in two-dimensional or three-dimensional form. For architectural design, the folder should contain the design work, and the made product should be a well-constructed architectural 3D model.

Models are not appropriate as made products for Resistant Materials and Systems and Control. For example, it is inappropriate to produce paper/card models as the final outcome for products that should be manufactured using resistant materials. You should create a product that can be properly tested and evaluated in the environment for which it is intended.

You should be able to demonstrate that you have been through all of the following stages in your project work:

1. identifying a need and establishing a design brief
2. researching the design brief and developing a specification
3. generating and exploring design ideas
4. developing a solution
5. planning for production
6. making a product
7. testing and evaluation.

When working through your project, follow this design process. It is important not to miss out any of the stages. In addition, try to use a range of presentation techniques throughout the project folder and demonstrate a range of ICT or computer skills in many different forms, such as graphs, charts, 2D and 3D design work, and CAM.

The following pages will give you information on what is required in each of the stages 1–7. This is your chance to show in practice what you have learnt and how skilled you are at designing and making a product.

1. IDENTIFYING A NEED AND ESTABLISHING A DESIGN BRIEF

Finding a suitable problem can sometimes be a challenging task. If it is too simple then you may find it hard to demonstrate and use all the skills you have learnt. If it is too complex then you may not address the problem sufficiently or even manage to complete the task. Remember that you will be spending many months on your project so finding a problem that you are interested in is extremely important too. Some things to consider when choosing your project include:

- Is there a product that could be improved for a hobby or interest you have?
- Ask your friends and relatives; they may have a few suggestions or ideas about design problems you can solve for them.
- Is there an item that you use every day that could be improved?
- Is there an item that could be developed for disabled people, young children or the elderly, or another exclusive group?

When you have identified your design need or opportunity, you will need to write a problem description. It's a good idea to state the following:

- What is the problem?
  Also add photographs or sketches of the problem.
- Where is the problem?
  Add photos or sketches.
- Why has this problem been identified?
  Give details of the circumstances for choosing the project.
- Who is the user of the intended product?
  Present details of your target market or customer profile.
- What is the intended finished product?
  Give details of what you are intending to make!

You can support your description with photographs and sketches to help explain your ideas and thinking. Make sure the text is clear. At this stage, there is no need to be too specific about materials, sizes and how the product is going to work. Keep statements and sentences general; you can add details later, when you have a full understanding of all the elements of the problem.

You will then need to produce a clear design brief, usually in paragraph form, for example:

Glossy Magazine Project

Situation

A well-renowned publishing house responsible for some of the **country's most popular magazines** wants to add a new title to its range. They have no preference over the target audience for the magazine, but require the magazine to be both **popular** and easily **marketable**. They will launch the magazine **in style** with a **free gift** and special promotional materials.

Design Brief

I will design and make a selection of samples for a new magazine. I will not be required to produce the whole magazine, but will have to produce a cover, sample pages and an appropriate three-dimensional free gift that could be given away with the first issue. I will need to:

- choose an appropriate target audience and magazine theme and name
- design and make a front cover for the first issue of the magazine
- design and make some sample articles for the magazine
- design and make a 3D product that would be given away free with the first issue
- design and make a suitable point of sale display stand for the launch of my new magazine.

---

2. RESEARCHING THE DESIGN BRIEF AND DEVELOPING A SPECIFICATION

Once you have a clear design brief you will need to analyse it. At this stage of the process it is a good idea to gather as much information as you can about the problem and situation. Ask lots of questions and record them.

One initial technique is to highlight the key words from the design brief statement and analyse them. Take a keyword such as *target audience*:

Analyse the word to help you work on understanding your brief. Do this for all keywords in your brief on one large mind map, such as the one above. Put this in your project folder as it demonstrates your thinking and analysis.

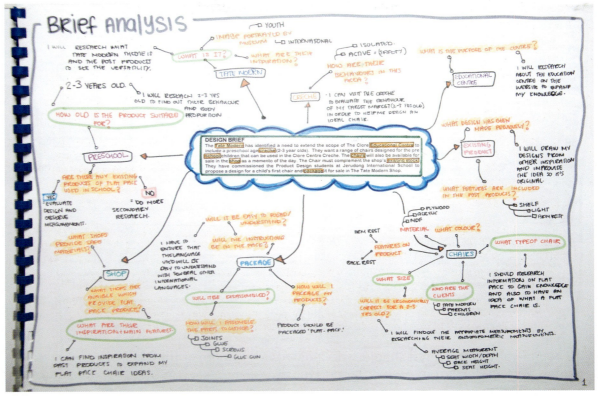

△ A student's analysis of a brief to design a chair suitable for children to be used and sold at the Tate Modern museum in London.

Another technique is a 'P's analysis. Look at the following words and relate them to your brief:

- people
- places
- processes
- products
- price
- pleasure.

Other research may include interviews, observations, surveys, and disassembly and analysis of similar products. When you have completed your research, it needs to be gathered, ordered, analysed and presented. Make sure the research is useful to your theme/brief. Ensure your brief meets the requirements of the user and discard any irrelevant information.

△ A student's product analysis of existing chairs. Looking at other products helps you to form a specification.

All sections of your research should be evaluated. Make note of the conclusions you draw from each different type of research, such as observation, data collection, user interviews and product analysis. At the end of each piece of research you could create a box on the portfolio page to write a summary of your conclusions. You could do this in bullet points. This will help you write a detailed specification later.

The specification

The specification is a list of design considerations your product will follow. If your research is thorough and detailed the specification should be easy to write as your conclusions for each aspect of the research should help you. Specifications are usually bullet point lists of statements. Keep these short and clear.

A SPECIFICATION:

- is a list of requirements
- sets out the key features of a product
- is a series of bullet points
- is clear and succinct
- is precise and specific using data wherever possible
- relates to the research
- is the criteria for evaluation
- is used throughout designing

◁ A summary of what makes a good specification. You should aim to cover all of these points in your own specification.

△ The specification is influenced by your initial brief analysis and all of your research.

3. GENERATING AND EXPLORING DESIGN IDEAS

This is the really fun and creative aspect of the design process. Try and use a range of presentation techniques here, whichever area you are specialising in, including:

- thick- and thin-line drawing technique
- marker renderings
- coloured pencil rendering
- use of shade and tone
- 3D isometric and perspective drawing
- 2D line drawings
- CAD drawings.

When designing, you should try to demonstrate a wide range of solutions with developments, no matter how impractical they are! This stage is all about ideas, creativity and innovation. You need to create an idea and develop it. Try to make each idea conceptually different. This is where the real thinking takes place.

It is a good idea to annotate your drawings so that they relate directly to the specification. You should try to include construction and minor details, decoration and any images or notes that have helped develop your ideas. Label and identify any technical aspects of your design, for example, how parts are joined, which materials are used, measurements and sizes. Make the annotation clear and relevant to your specification to help you justify your decision later on when you have to choose a specific idea to continue with and develop thoroughly.

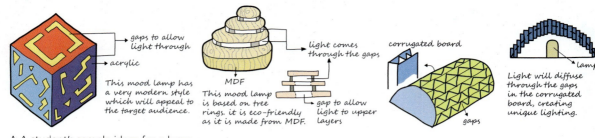

△ A student's sample ideas for a lamp

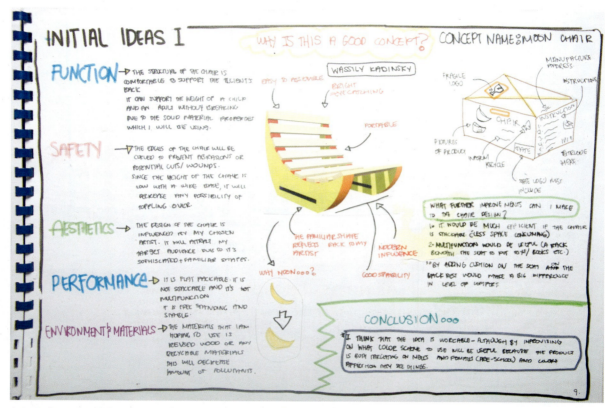

△ A student's initial design ideas for the children's chair. Remember to evaluate your ideas against the specification - do they fit?

4. DEVELOPING A PROPOSED SOLUTION

In this stage of the project you will need to bring together the best and most suitable ideas from the previous stages and make decisions about form, materials, construction, production methods, and finishes. You should try to include information about:

- material finishes
- fixing methods
- joints
- structural elements
- moving and mechanical parts
- electronics sub systems
- modelling the form of the product
- suitable working drawings:
  - orthographic
  - isometric

- electronic circuits
- mechanical diagrams
- perspective drawings.

It is also advisable at this stage to demonstrate that you have carried out appropriate and specific testing with regard to most of the points covered in your specification. For example, if you are designing an electronic system, have you tested a range of circuits for suitability and shown the changes you have made to get a working solution?

It is important here to develop the project. Avoid regurgitating lists of materials from a textbook that does not relate to what you are doing; a long list of wood finishes with images serves no purpose if it does not relate to the idea proposed, with a polymer finish. However, examples of wood finishes that you have tried, and you have discussed with your client or user, would be very relevant and appropriate.

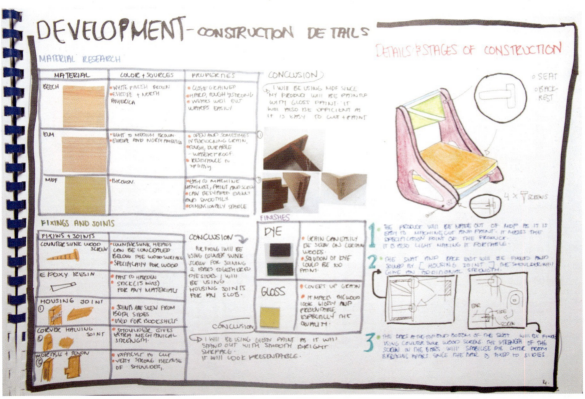

△ A student's development of their best idea for a chair suitable for children

5. PLANNING FOR PRODUCTION

At this stage, your project folder should have enough information in it for a third party to be able to make your project from the information given: a plan of making, working drawings, and a list of materials required. All working drawings should have the correct measurements and provide all the information required for the complete product. In the case of an electronic product this could include circuit diagrams as well as the case to house the project and the associated interface or controls. The plan of making should include all the steps to make the product, as well as consideration of quality checks (for accuracy) at various stages, and safety checks for safe-working practice in the workshop.

Assembly planning

Process outline	Tools needed	Step by step	Quality control	Quality check	Safety check	Comments
First gather all materials and measure and cut them into size	300 mm ruler, pencil, protractor, tri-square, copin saw, scroll saw, dovetail saw, bench hook	After collecting the adequate amount of wood and see-through acrylic rods, I have to label them to size and cut. Acrylic is cut with dovetail saw and wood is cut by band saw	Measure and cut wood and acrylic	Check measurements as I go along	Wear apron, goggles	I must not cut into precision so I have extra space to do my finishing (cutting label must be at least 5 or 3 mm from original line)
Briefly sand angled edges and rough surfaces of wood and acrylic rods	Sanding machine, ruler tri-square	Briefly sand the small labelled edges of wood parts that are to be cut. Sand uneven edges of poles	Check measurements consistently	Make sure the measurements that are to be cut are fulfilled	Wear goggles and apron	To check the rods, let rod stand on flat surface and use protractor to see whether it is 90 degrees straight
Label each part of wood piece on where to drill and with the acrylic poles on where to bend and drill	300 mm ruler, pencil, protractor, tri-square	Part must be labelled with precision	Check measurements consistently	Check measurements as I go along	For safety of product make sure that you are in a spacious area so no parts are dropped and damaged	

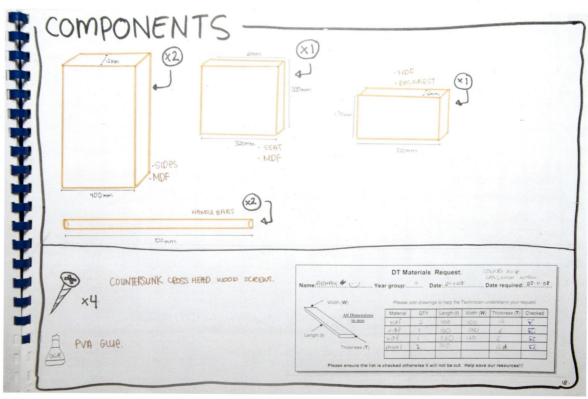

△ A list of component's for a student's design idea. Make sure to include correct measurements in your own project.

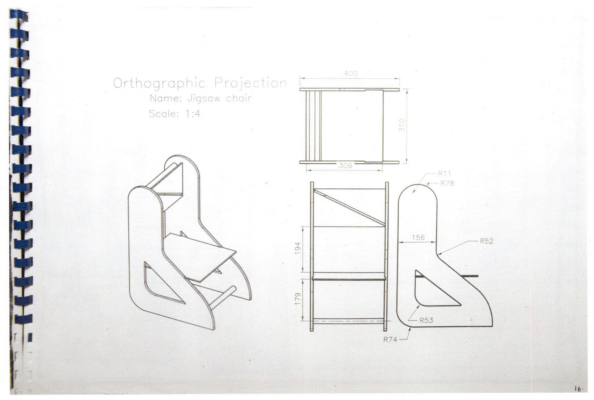

△ A student's first-angle orthographic projection drawings for parts of their project

6. PRODUCT REALISATION

Making a product is, for many, the highlight of the design process. Here is an opportunity for you to manipulate a range of materials and to produce usable, high-quality products for the target user, as set out in your design brief. Your final product should be based on your design drawings from the previous stages, but modifications can be made as the making process progresses. Designs often have to be modified at this stage as it is easy to overlook potential problems and constraints in the design stages. Do not be afraid of this and avoid thinking you are getting it wrong. However, make notes of any changes or modifications and refer to these in the evaluation stage of the designing process. It is often useful to use a range of construction techniques and different materials in order to demonstrate your abilities.

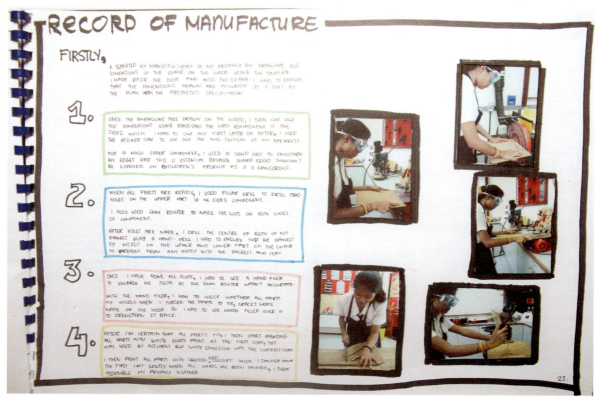

△ A student's record of manufacturing a chair

7. TESTING AND EVALUATION

Once you have produced your final product, it is a good idea to carry out testing to ensure it is safe and does what it is supposed to do. Consider the following:

- Has the product been tested in the environment proposed in the brief?
- Has the product been tested with the intended user group?
- Has the product been evaluated against the design specification?

When writing up this section you can use photographs, text and diagrams to illustrate your findings. Include feedback from people who have helped you test your product. Yours and others' opinions are extremely valuable and useful here. The table on the next page is a student's evaluation of a lamp:

Specifications		Additional	Further comments
The lamp has to be able to slide and adjust the amount of light that can be shown	🙂	When unscrewed and screwed again it does seem to get tight, not as easy to slide as usual, but this does then go away. This may be resolved in a more neater system of screws on the components	The lamp is able to do this quite well without much effort to slide the component, as well as not damage or affect the quality of the light coming out
The bulb has to be able to be easily accessible	🙁	This meant the lamp had to be easily opened to access the bulb for replacement	This could be done by making extra panels of acrylics and designing a design to have some sort of lid feature, which didn't need to use screws
Have a futuristic design	🙂	N/A	The lamp has a slight resemblance to the very futuristic ideas drawn by the architect Zaha Hadid
Needs to have multiple wooden components for a wooden theme	🙂	N/A	Most of the lamp is made from the pine wood material so this is achieved throughout the product
Sturdy build when in the open and closed position	🙁	This could be resolved by tighter spaces between all the components. Also perhaps using stronger materials	Not as rigid as the open formation. It does seem quite flimsy when in the open position
Mood light brightness of the lamp when turned on	🙂	This is because his preferences were a lamp that was in a mood light setting as well as nightlight	This was done nicely as the shade was covering use of the opaque acrylic, which has turned out to be an ideal material to use.

Finally, make proposals for further modifications and development where the design does not meet the design specification or customer requirements. You can do this in text format but drawing out your modifications is also very useful. Include comments about the strengths and weaknesses of the product and an overall evaluation of your work on the project. See examples from students below.

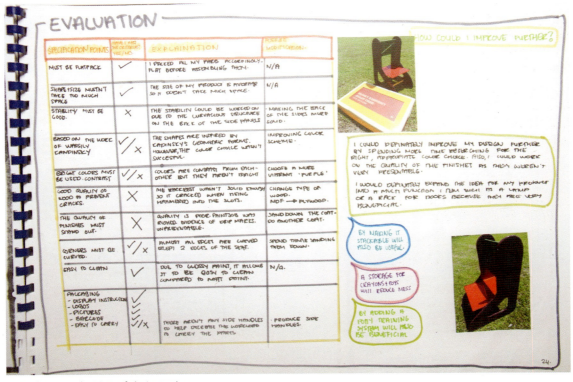

△ Student's evaluation of their product

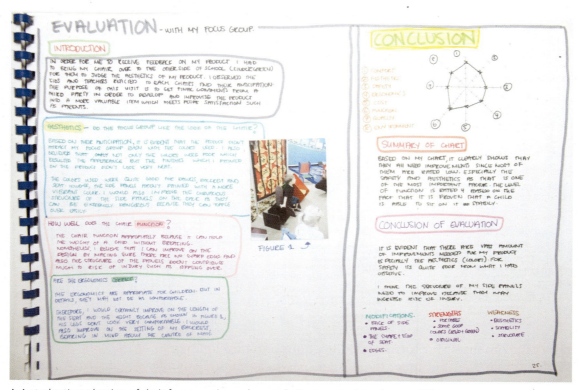

△ A student's evaluation of their focus group, and a conclusion summarising how well the product met the brief and how to improve it.

Glossary

555 integrated circuit A type of integrated circuit used for a variety of timing applications.

addition The process of shaping materials by combining or joining them, for instance using screws, nails, nuts and bolts, and adhesives. Joining methods can be classified as temporary or permanent, with particular methods being used with different materials.

alloy A metallic material that is a mixture of two metals.

anthropometrics data Information about the measurements and properties of the human body (e.g. hand sizes of 15-year-olds).

applied load The size of a load acting vertically on a beam.

artisan A highly skilled person who generally produces bespoke or small numbers of high quality products or items.

assembly drawing An isometric view showing parts and components assembled.

astable output An output that changes between being on and off while the circuit is operating.

base One of three pins of a transistor, used to control the switching or amplification action of the transistor.

battery An electrical device that can store energy.

beam A structural member which spans gaps and supports the loads acting on it.

bearing A mechanical device used to reduce friction between moving parts.

bespoke Made to the specifications of an individual customer, usually resulting in an original and unique one-off item.

biomimicry An approach to innovation and design that takes inspiration from how nature solves problems. Materials developed using biomimicry include polymers with adhesive properties inspired by a gecko's feet, and fabrics with streamlining properties inspired by shark skin.

bitmap graphic A graphic that is made up of tiny dots called pixels.

blister packaging A type of packaging with a moulded plastic cavity or pocket.

brainstorming The process of pulling apart an idea, theme or problem in order to analyse it, discuss it and to generate new ideas.

British Standards The commonly accepted way in the UK of drawing lines and adding dimensions.

buckling A type of failure in which a component bends out of shape, usually as the result of pressure or heat.

cam A specially shaped part attached to a rotating shaft that imparts linear motion to a cam follower.

capacitor A two terminal component used to store an electric charge temporarily.

client The person or company you are working for to create a solution.

closed-loop system A system which incorporates feedback to control and manage the system.

CMOS logic gate family A category of logic gates (4000 series) commonly used for school projects.

collector One of three pins of a transistor, normally connected to a positive part of a circuit.

column A pillar designed to support the weight of a structure pressing down on it.

commercial testing Detailed and controlled testing carried out in laboratory conditions.

comparator A circuit which is designed to compare two input voltages and is often used for sensing purposes.

composite A material that is made from two or more materials that have significantly different physical or chemical properties. When combined, they create a single material which is often lighter and stronger than the original materials on their own.

composite sandwich beam A beam constructed by a lightweight material sandwiched between two stronger materials.

Computer-aided Design (CAD) The use of computers to design products on a screen. Also used in production management.

Computer-aided Manufacture (CAM) The use of computers to enable the manufacture of products and to control production processes.

Computer-integrated Manufacturing (CIM) A method of production in which design, manufacture and management are linked electronically and controlled by computers.

constraint Something that may limit aspects of your design.

Control of Substances Hazardous to Health (COSHH) A set of regulations designed to ensure the safe control of hazardous substances.

costs The amount of money needed to manufacture a product, including materials, tooling and labour.

cradle to cradle The whole life of a product from raw materials to end-of-life and recycling.

crank and slider A mechanism used to change rotary motion to linear motion, or vice versa.

crating The practice of drawing inside a cuboid to help the drawing of isometric or planometric projections.

cross-sectional shape A shape made by cutting through something, for instance ties, struts, columns and beams.

cure The process of allowing a substance, for instance a polymer resin used within a composite, to harden and set.

cutting plane An imaginary plane that cuts an object.

datum A point used as a reference point for all measurements when marking out material.

deforming The process of subjecting a material to a stress that changes its shape. Methods include bending and the use of jigs and formers. Typically heat is applied to materials to bring them into their 'elastic' zone, where they can be deformed.

design brief A statement of your design intent or a description of what the product will be like.

design process A logical problem-solving process of designing a product.

development/net A folded-out net of a container, packaging, or a three-dimensional shape, e.g. a piece of packaging that has been flattened and folded out.

die cutting/die stamping The commercial process of cutting material using a shaped metal die.

dimensions The measurements of the size of an object, usually shown in mm.

diode An electronic component which allows current to flow predominantly in one direction only.

disassembly The taking apart of a product in order to look carefully at how it is made, the materials used and processes involved in manufacture.

discrete components Separate individual components such as resistors, capacitors, transistors and diodes.

driven gear or pulley A gear or pulley driven by a driver gear or pulley.

driver gear or pulley A gear or pulley used to drive the system it is connected to.

efficiency The ratio of useful work or power to actual input work or power.

effort The force applied to the input part of a mechanism.

elastic state The state in which a thermoforming polymer is at a temperature that produces a stretchy consistency and in which it can be shaped, but will not retain its form without being held in place. The elastic state is used, for instance, in line bending, vacuum forming or dome blowing.

electrical current A flow of electricity through a circuit or a measure of the rate of this flow – unit of measurement: Amp (A).

electrical solder An alloy with a low melting point, used to join electrical connections.

ellipse An oval shape, like a flattened circle. The larger diameter is the major axis and smaller diameter the minor axis. It is also a shape produced when a cone is cut through at an angle.

emitter One of three pins of a transistor, normally connected to a negative part of a circuit.

end user Someone who uses or will use your product.

ergonomics The study of the way products can be designed so that they produce the best experience for users.

exploded drawing An isometric view showing parts and components separated.

exterior Suitable for outside or outdoor use.

facilities The equipment, processes and machines available to be used in the design or manufacture of a product.

factor of safety The ratio of the ultimate breaking strength of a material against actual force exerted on it.

farad The unit for measuring capacitance (the ability to store electricity).

feedback Information and opinions about the work or product from people who have seen or used it

or The use of sensors to read the output and tell the input what to do.

or Part of a closed-loop system which monitors the output and sends a signal back to the input. This signal may cause an adjustment (if necessary), in order to achieve the target output level.

ferrous A metal that contains iron and which will rust unless protected, for example by being painted.

field test *or* **user trial** A semi-controlled test by intended users of a prototype or finished product.

finish A substance that is used to produce the protective layer or texture colour for the surface of a product to make it last and look good.

first-angle projection A method of laying out an orthographic drawing with the end elevation in the top left and the plan view below the front view.

fixtures and fittings Standard components used to assemble a product that have been purchased rather than specially manufactured.

flaps Parts of a development that stick out, used for tucking sections (such as lids) inside securely.

Flexible Manufacturing Systems (FMS) Manufacturing systems that produce products and goods using robots and computer-controlled tools and machines without human intervention. The products can be easily changed by using a different program.

flowchart A diagram showing a sequence of operations in a process or work flow.

force A power that produces strain on a material and that changes its shape or has the potential to change its shape. There are five forces that act on materials: compression, tension, torsion, sheer and bending.

form The physical shape and appearance of something.

formative evaluation Evaluation that is carried out continuously during the development of a product.

forward bias The main direction of current flow through a diode.

frame structure A structure that is constructed using a 'skeleton' of individual parts.

fulcrum The pivot point about which a lever turns.

function The job that a product is designed to do.

gear system A system of toothed wheels which connect to transmit motion between shafts.

generator A machine that converts mechanical energy into electrical energy.

hardwood Wood that comes from broad-leaved trees, such as oak, beech and ash, that lose their leaves in autumn.

hidden detail Part or parts of an object that cannot be seen, shown by a dotted line on the drawing.

human factors The ways in which human users interact with products or systems.

hydraulic system A system which operates using oil or water.

I-beam A girder beam which has an 'I'-shaped cross-section with two flanges at the top and bottom of the beam.

input The signal that starts a system.

integrated circuit An extremely small circuit incorporating many components, etched onto a piece of semi-conducting material.

interaction design The aspect of a design that considers the relationship between hardware, software and the needs of the product's user.

interior Suitable for inside or indoor use only.

irregular polygon A shape with three or more straight edges of different lengths and with different internal angles.

isometric Describes a method of drawing a three-dimensional object at 60° and 30° angles.

jigs (*also* **templates**, **patterns** and **moulds**) Simple devices that speed up marking out and repetitive tasks.

kerning The process of adjusting the space between pairs of letters in a word.

key A list of explanations for the different categories used on a chart or graph.

kilowatt-hour A measure of electrical energy consumption equal to the kilowatts used by an electrical appliance for every hour it is switched on.

knock-down fittings Fittings that can be easily put together, normally using only a screwdriver, drill, hammer or other basic tools. They are temporary joints, although many are used to permanently join together items such as cabinets and other pieces of furniture that are purchased in a flat pack.

lever A mechanism used to move loads which will rotate around a fulcrum.

life cycle All the stages of a product's existence from its design and creation, through its use, to the end of its life.

life cycle analysis An analysis of each part of the product's life and journey, from creation to the end of the product's life.

life-size model A model that is the same size of the intended outcome.

light dependent resistor (LDR) A type of resistor that changes its resistance in response to the amount of light falling on it.

light emitting diode (LED) A type of diode which emits light when operating.

light source The point or place from which the light is coming.

linear motion Motion that follows a straight line.

linkage A set of rigid parts joining two or more components of a mechanism together. Each link has at least two nodes or attachment points.

load The force that is supported or moved by a mechanism.

logic gate An on/off switch that is the basic building block for electronic logic circuits.

lubrication The use of an oily or greasy substance to reduce friction between moving parts.

mains supply A general purpose electrical supply for buildings, supplied through cables.

manufactured board Board such as chipboard or hardboard that is made in specific sizes and with specific properties, made by joining or compressing small pieces of wood, often offcuts and waste from timber processing.

market form Standard forms and sizes of commercially manufactured material.

marking out The process of transferring measurements in the form of lines or shapes onto a material in advance of cutting, shaping, bending or forming.

mass structure A structure that is constructed by building or piling up materials such as stone, brick, concrete or soil.

matrix The material in a composite that performs the binding function.

mechanical advantage The advantage gained by using a tool, mechanism or machine.

mechanism A device which uses moving parts to perform a function.

memory The property by which a material returns to its original shape under specific environmental conditions, such as ambient temperature or heat from an electrical current.

micro-switch A highly sensitive small switch used in many systems and control applications (often operated by a level or roller).

moment A turning force around a pivot.

moment of force The magnitude of a force multiplied by its distance from the pivot point, expressed as the unit Nm.

monostable output An output that is on for a specified length of time while the circuit is operating.

nanomaterials A material manufactured at the molecular level to provide specific properties.

nanoparticles Particles that are between 1 and 1000 nanometers (10-9), and materials where the structure has been engineered at the nanoscale can have unique properties. For example, graphene (comprised of carbon sheets with the thickness of a single atom) has over 200 times the strength of steel, gram for gram.

natural structure A structure that occurs in nature and which is created by animals, birds, fauna and flora, insects and geological features.

Newton The unit of force which has the symbol N.

nominal resistance value A measure of value that states the resistance of a particular resistor.

non-ferrous A metal that does not contain iron and therefore will not rust.

non-polarised capacitor A small capacitor whose two connections can be connected either way round in a circuit.

normally closed (NC) A switch state where the switch is on until it is operated.

normally open (NO) A switch state where the switch is off until it is operated.

NPN transistor An NPN transistor is one of two transistor families (the other being a PNP transistor). NPN stands for negative-positive-negative and refers to the way the transistor is made in its silicon form.

off-shoring The movement of a company's operations, such as manufacturing or commerce, to another country, often to save costs.

open-loop system A system which is controlled without the benefit of feedback.

operation The 'brain' of a system. It takes an input, decides what to do and sends an appropriate message to the output.

operational amplifier A type of integrated circuit used to amplify low voltage input signals.

orthographic drawing A method of drawing a three-dimensional object in two dimensions, usually showing a plan view, front elevation and end elevation.

orthographic projection A two-dimensional drawing that represents a three-dimensional object, showing several elevations and dimensions.

oscillating motion Motion backwards and forwards or from side to side.

outcome The final project or product.

output A response to the information from the process element of the system.

out-sourcing The use of a third party contractor to carry out work and supply services.

packaging net A two-dimensional piece of material such as cardboard, designed to be folded and fastened into a three-dimensional piece of packaging.

PAR (Planed All Round) The market form of wood with planed edges. The actual size of PAR timber is smaller than the stated size owing to planing. A 50 mm × 25 mm PAR will therefore measure about 45 mm × 20 mm.

PCB circuit layout A line drawing that represents the conductive tracks that join components in an electronic circuit.

permanent joining A join that is not required to be separated and cannot be taken apart.

perpendicular Lines or surfaces that are at right angles (90°) to each other.

Personal Protective Equipment (PPE) Clothing and equipment, such as goggles or helmets, designed to protect a person working in a making environment.

physical property An observable or measurable characteristic of a material, for instance strength or density.

PIC ICs (programmable interface controllers) An integrated circuit that can be programmed to carry out a task or range of tasks.

planned obsolescence The deliberate creation of a product with a limited life.

planometric A method of drawing a three-dimensional object at 45° and 45° angles.

plastic state The state in which a thermoforming polymer is at a temperature that produces a consistency that is softer and more malleable than the elastic state. The plastic state is used, for instance, in injection moulding.

pneumatic system A system which operates using compressed air.

point load A load which is applied to a particular point on a beam.

point of sale display A board or other form of visual advertising that is displayed at the place where a product is purchased.

polarised capacitor A capacitor that has positive and negative connections and which must be connected the correct way in a circuit.

polymer A substance comprised of large molecules made up of simple molecules of the same kind. Plastics such as acrylic and polythene are the most commonly recognised polymers, but polymers exist in nature; for example, silk, wool and cellulose.

polymorph A smart material that can be heated in water to 62° and shaped. Once cooled it can be worked like a traditional material, with the advantage that it can be reheated and reshaped.

potential difference The difference in voltage across components, conductors and other sources when an electrical current flows through them – unit of measurement: Volt (V).

preparation The process of making the surface of the material ready before applying a finish.

primary function The main purpose of what a product is designed to do.

primary research New research that you have carried out yourself.

printed circuit board (PCB) A board with components mounted on etched tracks to form an electrical or electronic circuit.

printing plate A sheet which is wrapped round a roller and used to transfer the image to the substrate.

process *or* **control** The 'brain' of a system. It takes an input, decides what to do and sends an appropriate message to the output.

process colours The four colours that are used in printing: Cyan, Magenta, Yellow and Key (black).

prototype A working model in the early stages of development, which, through testing against the specification, is used to help refine the product before beginning final or larger scale production.

prototype board A thin, plastic board used to hold electrical components temporarily during the development of a circuit.

pulley A wheel with a grooved rim in which a belt runs.

qualitative data Information in the form of opinions.

quantitative data Information that can be easily expressed and analysed in a numerical form.

rapid prototyping Use of CAM to quickly produce models or parts for testing and modelling.

reaction force The force exerted by the beam supports which act to resist the applied load.

reciprocating motion A repetitive motion backwards and forwards or up and down.

reforming The process of changing the shape of a material, typically by melting and pouring or injecting the molten material into a mould, for example when injection moulding plastics or sand casting non-ferrous metals.

regular polygon A shape with three or more straight edges of the same length and with equal internal angles.

reinforced concrete A composite material that uses steel to strengthen concrete for structural use.

reinforcement The material in a composite that provides structure and adds strength.

relay A special kind of switch often used to control a high current circuit using a low current input.

rendering Adding line tone, colour and texture to a drawing to make it look realistic.

research The process of gathering new information about something.

resistance A material's opposition to the electrical current flowing through it – unit of measurement: Ohm (Ω).

resistant material A material that requires force in order to be worked or shaped.

resistor An electrical component designed to introduce a known value of resistance into a circuit.

resistor tolerance A measure of the permitted maximum and minimum resistance values of a particular resistor.

resolution The quality of images measured in dpi (dots per inch). The higher the dpi the better the quality of the print, but the larger the file size used to store the image.

resultant A single force that has the same effect as all the other forces acting on an object.

reverse bias The direction where there is no current flow until the diode breaks down.

risk assessment An examination of the possible risks in an area and a statement of how they should be managed.

rotary motion Motion in a circular path.

scale The ratio between the size of something real and a model or drawing of it, that has either been enlarged or reduced.

scale model A model that is proportional in size to the final outcome.

screw jack A device designed to lift a load, operated by turning a leadscrew.

seasoning The process of drying out timber so that it becomes strong and will not change its shape over time.

secondary function Another function that a product has in addition to its main (primary) function.

secondary research Research based on other people's work.

sectional view A view of a face of an object as if it has been cut through. Different types of sectional views include full, part, revolved and removed sectional views.

self-finishing A material that does not require a finish.

shell structure An external structure that has space inside.

sketch model or **mock-up** A quick, simple model of a design.

skills The abilities of the workforce or designer that need special training or expertise.

smart finish A coating that is applied to the surface of a material to enhance its performance, for example waterproofing or flame retardant spray.

softwood Wood that comes from fast-growing evergreen trees with cones and needles.

solar cell A device which converts light to electrical energy.

space frame A three-dimensional lightweight truss structure used to span large areas.

special finish A protective coating that is applied to the surface of a material.

specification A list of key points describing the construction, materials and appearance etc. that your design must have.

standard components A pre-prepared part that is used in the production of many products.

stiffness Resistance to forces of bending or torsion.

strain The amount of extension or compression of a material divided by its original length.

stress The force acting on a material divided by its cross-sectional area.

stress or **strain** A force created by pressure on an object or material, that has the potential to damage it by changing its shape.

stress-strain graph A graph showing the relationship between stress and strain for a material under load.

strip board A type of board used in electronic circuit construction which has a regular grid of holes positioned along parallel strips of copper.

structural member A single component of a structure.

strut A structural member designed to resist compressive forces.

substrate The material that something is printed on.

summative evaluation Evaluation that is carried out on a finished product, usually resulting in a detailed report.

surface finish A coating that is applied to the surface of a material in order to enhance its appearance.

surveys A set of questions asked to the public or to user groups in order to gather specific information to help develop products and services.

sustainability The ability of materials to be used wisely with little or no negative impact on the environment.

switch A device used to make or break an electronic or electrical circuit.

system A set of parts or processes which work together to achieve an output or perform a function.

system diagram A diagram that shows the main parts or functions of a system, usually in the form of a block diagram where they are represented by blocks joined with lines.

tabs Parts of a development that stick out, and which allow sections of it to be glued together.

tangent A line that touches a circle or curve at one point but does not intersect it.

target market The group of people you hope will use your product.

thermistor A type of resistor in which the level of resistance depends on its temperature.

thermoforming Plastic materials that can be repeatedly softened by heat and formed into shapes, which become hard when cooled.

thermosetting Plastic materials that can be softened and formed into shapes which become hard when cooled and cannot be softened again.

third-angle projection A method of laying out an orthographic drawing with the end elevation in the bottom right and the plan view and front elevation to the left.

tie A structural member designed to resist tension forces.

timber Wood that has been processed and turned into a form that can be easily used, often planks.

tolerance The acceptable margin of error in a measurement when marking out materials ready for cutting and shaping.

tracking The process of adjusting the space between all the letters in a word.

transistor A small three-contact electrical component used for fast switching or amplification.

triangulation The use of triangular shapes to give stability to a structure.

truncation A cylinder or prism that has a part of it sliced away.

truss A structural framework used to support loads, for example in a roof or bridge.

two-point perspective A method of making a realistic three-dimensional drawing using two vanishing points and a horizon line.

typography The art of designing, placing and adjusting lettering to make it readable, legible and attractive for printing.

uniformly distributed load A load which is distributed equally along the length of the beam.

user group The type of people who use or will use your product.

vacuum forming The process of heating up a plastic sheet and using a vacuum to suck it over a mould.

vector graphic A graphic that is created through mathematical information.

velocity ratio The ratio of distances moved by the effort and load at the same time.

veneer A thin layer of a substance (often timber) that is applied as a surface coating to something, often to increase the aesthetic appeal of the underlying material.

voltage A measure of electrical current, measured in volts – unit of measurement: Volt (V).

warping The distortion or twisting that can occur to timber, often as a result of poor storage, poor seasoning or natural defect.

wastage The process of cutting away material to leave the desired shape, for instance using saws, files and abrasives.

wheel and axle A simple machine part that consists of a wheel attached to an axle and which is used to lift loads or transmit power from one part of a mechanism to another.

working property The way a material behaves when being worked or shaped, or while being used within a product. Working properties determine the tools or processes that will be used when making or manufacturing, and how a product will function.

x-axis The horizontal line of a graph.

y-axis The vertical line of a graph.

Index

3D printing 28, 65, 66, 100
555-timer integrated circuit 331–3, 334
741 op-amp 339, 340, 342

A

ACCESSFM (aesthetics, customer, cost, ergonomics, size, safety, function, manufacture) 13, 53
accidents at work 55–6
acetate sheets 152
acrylic (PMMA) 45, 173, 174, 220
acrylonitrile-butadiene-styrene (ABS) 174, 176
addition 205, 209, 210
adhesives 44, 152–3, 175, 220, 255
Adobe
　Illustrator 143, 157
　Photoshop 143, 156
aesthetics 92–3
air bearings 296
alloy 186, 189, 190, 191–2, 195
　shape memory alloys 152, 170, 171
alternating current (AC) 299
aluminium 191, 192, 193, 210, 223
American National Standards Institute (ANSI) 88, 111
AND gate 333, 335, 337, 338, 339
annealing 192–3
anthropometrics data 12, 15, 17, 75–6, 77
applied loads 245–7
appraisal, of design ideas 30–1
artisan 72, 77
ash 179, 183, 209
assembly drawing 117–18, 119
astable output 331, 332, 333, 334
attribute-analysis diagrams 27
AutoDesk® 23, 64

B

ball bearings 295, 297, 299
bar charts 144
barcodes 67
base 320, 322
batch production 65, 80
batteries 292–3, 294, 299, 306
beams 237–8, 239, 244, 253
　bending 243
　non-symmetrical 251–2
　uniformly distributed load 246–7
bearings
　definition 295, 298
　selecting 294, 296–7
　types of 295–6
beech 179, 183, 209
belt drives 286–8
　belt types 286, 287–8
　tensioning 288–9
bespoke production 64, 80, 82
bevel gears 284
bicycles 235, 265
biomimicry 84, 170, 172
bitmap graphics 156–7, 158
blister packaging 160, 162, 163
blockboard 180
blow moulding 43, 162, 173
brainstorming 9, 13, 14
brass 191, 210, 211–12
brazing 211–12, 255, 256
breadboard *see* prototype board
bridges 237, 241
British Standards (BS) 111, 114
British Standards Institute (BSI) 60, 61
brittleness 189, 190, 193
bronze 186, 191
buckling 243, 244

C

calipers 197, 202–3
cam lock 215, 216–17
cams 275–7, 280
cantilever beam 237, 253
capacitors 328–9, 330–1
carbon-fibre reinforced polymers (CFRP) 185, 187, 235
carcases, construction 215, 221
card 26, 43, 151
case hardening 194
casting 44
catches, locks and latches 219–20
CE marking 61, 88
charts 38, 40, 87, 144–7
chipboard 46, 180, 183
chisels 206, 208
circles
　drawing 124
　enlarging 130–1
circuit diagrams 305, 306, 307, 308, 332
class 1 levers 266, 267
class 2 levers 234, 244, 266, 267
class 3 levers 266, 267
client 8, 9–10, 30, 72
closed-loop systems 231–2
CMOS (Complementary Metal Oxide Semiconductor) 336, 343
collector 320, 322
columns 239
commercial testing 52, 54
communication
　communicating design ideas 32–5
comparator 342, 343
compasses 42, 113, 123, 126, 130–1, 134, 202
composite materials 184–8
composite sandwich beams 243, 244
compression 255
computer-aided design (CAD) 68, 158
　developments/nets 64, 127
　graphs and charts 145
　light sources, showing 139
　in manufacturing 63–4, 65
　modelling 28, 99, 100
　Project work 344, 346
　rendering 23, 24, 33, 64
　software 23, 64, 157
computer-aided manufacture (CAM) 68
　applying text 143
　in manufacturing 66
　modelling 27, 28, 100, 152
　Project work 344, 346
　shaping materials 208
　types of 66
　vector designs 157
Computer Integrated Manufacturing (CIM) 66, 68
Computer Numerical Control (CNC) machines 66, 100, 208, 289

concrete 184–5
    reinforced concrete 185, 186, 233, 242, 244, 251, 252
constraints 9, 12, 13, 16
    identifying 86–7
construction kits 28, 37
construction lines, drawings 117, 137, 138
continual improvement (CI) 54
continuous production 65, 81
Control Of Substances Hazardous to Health (COSHH) 59, 60, 62
control systems 69–71, 230, 344, 346
    logic control systems 338–9
conventional current flow 299, 304
copper 186, 191, 192, 223
CorelDRAW 64, 157
corner blocks 46, 216
corrosion 190, 191, 222, 223
corrugated card and plastic 151
costs 36, 40
countersunk hole 208, 217
cradle to cradle 73, 77
craft knife 159, 162
crank and slider mechanisms 274–5, 280
crating 112, 114, 118
creasing and folding tools 160
cross-bracing 240
cross-sectional shape 242, 243, 244
cure 185, 188
cutting plane 115–16, 119
cutting tools 43, 159–60, 162, 197, 206
cylinders 289–90

D

data analysis, interpretation and recording 18–19
data graphics 144–7
datum 41, 47, 197, 199, 203–4
deforming 205, 209, 210
design & technology in society
    designer, role of 72–3
    effects of D & T activity 73–4
    human factors in design 75–7, 78, 79
design brief 8, 10, 11–12, 25, 30, 72
    establishing 347–8
    researching 348–50

design movements 94–6
design need 8–9, 347–8
design portfolios 17, 27, 35, 344
design process 8, 10
    constraints on 9, 12, 13, 16, 86–7
    evaluation 27, 29, 30–1, 48–54, 351–2
    modelling *see* modelling
    printed circuit boards 306–8, 330
    research and 14–15, 16, 17, 72–3
    SCAMPER (substitute, combine, adapt, modify, purpose, eliminate, rearrange) strategy 21–2, 84–5
    selecting mechanisms for 291
    use of technology in 63–8
designers
    energy conservation and 104
    role in society 72–3
development/net 64, 68, 127–9, 162
diagrams, planning 87
dial gauges 253
die cutting/die stamping machines 43, 127, 160, 163
dimensions, drawings 23, 25, 111
diodes 321, 323–5
direct current (DC) 293, 299
disassembly 48–50, 54, 105
discrete components 321, 330, 346
dome blowing 173, 176
Double Pole Double Throw (DPDT) relay 313–14
drawing
    2D and 3D drawings 23, 33, 63, 66, 97
    assembly drawing 117–18, 119
    basic shapes 122–6
    enlarging and reducing shapes 130–3
    formal techniques 110–14
    freehand 20, 22–4, 32, 120–1
    graphs and charts 144–7
    instruments and drafting aids 134–5, 197
    layout and planning 136–8
    presentation 139–43, 351–2
    scale drawings 97, 151
    working drawings 23, 33, 352–3

see also exploded drawing; isometric drawing; orthographic projection
drawing boards 134
drills 206, 207, 208
driven gear or pulley 281–2, 284, 286, 287, 291
driver gear or pulley 281–2, 284, 286, 287, 291
ductility 167, 189, 190, 191

E

eccentrics 275, 276
economic design issues 74
efficiency 271, 272
effort 266–8, 272
elastic state 173, 176
elasticity 167, 189
electrical current 307–8
    measuring 303–4, 304
electrical solder 212, 309, 310
electronic circuits 299
    integrated circuits 307, 331–4, 342, 343
    printed circuit boards 305–8, 310
    time delay circuits 330–4
electronic kits 28
electronics
    basic concepts 299–304
    capacitors 328–9, 330–1
    circuit building techniques 305–10
    diodes 321, 323–5
    electrical power, calculating 300–1
    integrated circuits 307, 331–4, 342, 343
    logic gates 333–4, 335–9, 343
    operational amplifiers 339–43
    resistors 316–19, 324, 326, 327, 332–3, 338–9
    switches 311–15, 321, 338–9
    time delay circuits 330–4
    transducers 326–7
    transistors 320–2, 326
ellipses 124–5, 126
emitter 320, 322
end user 8, 9–10, 30
    meeting needs of 78–9
energy
    costing 293–4
    energy conservation and efficiency 103–4, 294
    energy sources 101–2, 292–3, 294

enlarging shapes 130–3
environment
    design issues 74
        energy use and 101–4
        timber and 182–3
epoxy resin 175, 220
ergonomes 78, 97
ergonomics 75, 77, 78
evaluation in design
    definition 48
    evaluating against the
        specification 50–1, 89, 356–7
    evaluating existing products 16,
        17, 48–50, 84–5
    of ideas 27, 29, 30–1, 48–54,
        351–2
    modifications based on 53–4,
        357–8
    product testing and evaluation
        52–3, 356–8
exclusive design 98
exploded drawing 33, 117–18,
    119, 121
exterior 222, 224, 227

F

facilities 40
factor of safety 262, 264
farad 328, 329
fastenings 44, 217–20
    see also adhesives
feedback, opinions 9, 10
feedback, system output 70–1,
    231, 232
ferrous metals 189, 190, 195
field test 52, 54
finishes 17
    choosing 15, 43, 225
    health and safety 226
    importance of 222
    self-finishing materials 222,
        225, 227
    types of 222–5
first-angle projection 110, 114,
    355
fixed resistors 316, 317, 324
fixtures and fittings 36, 40
flaps
    developments/nets 128–9
    joining materials 153–5
flat-packed furniture 117, 215
Flexible Manufacturing Systems
    (FMS) 68
flexicurves 134
flexography printing 148
flowcharts 38, 40, 87, 146–7

fonts 142, 156
force 169, 258–62
    determining forces 262–4
form 83, 85, 92
form follows function 91
formative evaluation 48, 54
formers 65, 182, 209, 210, 211
forward bias 323, 325
fossil fuels 101–2
frame structures 234, 236
    strengthening 240–1
    types of 237–9
frames, construction 221
framework structures 234, 239
freehand drawing 20, 22–4, 32,
    120–1
fulcrum 266–8, 272
function 83, 85, 90
    form follows function 91

G

gain, transistor 320, 321, 340–1
gear ratio 270–1, 282, 285
gears and gear systems 291
    simple and compound gear
        trains 281–3
    and torque 283–4
    types of 284–6
    velocity ratio 270–1, 282–4, 285
generators 293, 294, 299
glass-fibre reinforced polymers
    (GRP) 185
graphic design 108, 142, 156
    modelling, importance of
        150–1
    presentation in 139
graphic products 108, 344, 346
    manufacture of 159–63
graphs and charts 144–7
gravure printing 148
gusset plates 240

H

handmade models 99
hardboard 180, 183
hardening steel 193–4
hardness 167, 175, 189
hardwoods 177, 179, 183
health and safety 226
    accidents at work 55–6
    in Design & Technology
        areas 60
    designers and 60–2
    importance of 55
    Personal Protective Equipment
        (PPE) 57–8, 62, 186, 225–6

regulation and guidance 58–9
    risk assessment 56–7, 62
    soldering 310
    workshop symbols and signage
        59–60
helical gears 284
hidden detail, drawings 110, 114
high density polyethylene
    (HDPE) 174, 176
hinges 219
horizon line, drawing 113, 114, 120
human factors in design 75–7, 78
hydraulic systems 289, 291

I

I-beam 243, 244
ICT, graphic design and 156–9
ideas
    communicating design ideas
        32–5
    evaluating 27, 29, 30–1, 48–54,
        88–9, 351–2
    generating 20–4, 30–1, 83–5,
        351–2
    initiating and developing 18–19,
        352–3
IKEA 39, 46
images
    creating your own 156
    manipulating with CAD 156–7
inclusive design 98
injection moulding 43, 173, 174,
    176, 210
input 69–71, 230–1
integrated circuit logic gate
    families 336–7
integrated circuits 307, 331–4,
    342, 343
interaction design 77, 78, 79
interior 224, 227
International Organization for
    Standardization (ISO) 88, 111
inverting amplifier 341–2
irregular polygons 124, 126, 131
isometric drawing 32, 112, 114
    freehand 120, 121
    laying out 137
Ive, Jonathan 96

J

jigs 65, 68, 80, 198, 210, 211
joints
    metals, joining 44, 45, 211–14,
        217, 218, 256–7
    modelling, joining methods
        152–5

permanent joining 154–5, 213, 255, 256
plastics, joining 44, 45
rigid and moveable 255
in structures 255–7
timber, joining 44, 182, 211, 215–17, 217–19, 221

K
kerning 143
Kevlar® 186
key, graphs and charts 144–5, 147
kilo-watt hour (kWh) 293, 294
kilonewton (kN) 255
knock-down (KD) fittings 44, 45–6, 47, 211, 215–16

L
laminating 43, 182, 205, 209, 251
laser cutters 66
latch circuit 313, 314–15
laying out drawings 136–8
Lego® 26, 28, 39
lenticular sheet 171
lettering and fonts 142–3
levers 234, 244, 248, 266–8, 272
life cycle 10
life cycle analysis 73, 77
life-size model 26, 29
light dependent resistor 326, 327, 339
light emitting diode (LED) 321, 324–5
light source 139, 140, 143
line bending 173, 176
line graphs 144
linear motion 273, 280
  rotary motion, conversion 265, 273–7, 280, 281, 285–6
linkages 268, 269, 272
load 266–8, 272
  applied loads 245–7
logic control systems 338–9
logic gates 333–4, 335–9, 343
low density polyethylene (LDPE) 174, 176
lubrication 271, 294, 296, 297, 298

M
machine screws 217, 257
magnetic bearings 296
mahogany 179, 183, 224
mains supply 292, 294

malleability 167, 176, 190, 191, 194
manufactured boards 46, 180–1, 183
manufacturing
  CAD/CAM in 63–4, 65, 66
  graphic products 159–63
  implementation and realisation 41–7
  materials selection 36–7
  planning for 37–9, 353–5
  production systems 64–6, 80–2
  Project work 353–6
  selecting processes for 42–4
  tolerance 46, 47
  use of technology in 63–8
manufacturing plan 38–9, 354
market form 196, 198, 199
marketing mix 12–13
marking out 41–2, 200–2, 203, 204
mass production 65, 81
mass structures 233, 236
materials 164
  classification of 166
  composite 184–8
  cutting methods 43, 197
  finishes 222–7
  joining and assembly methods 44, 45, 152–5, 182, 211–21, 256–7
  for manufacturing 36
  materials list 37, 353, 354
  for modelling 26–7, 151–2, 168
  preparation of 196–9
  properties of 166–7, 169, 251–2
  recycling 106, 176
  rendering 141–2
  resistant materials 210, 344, 346
  selecting 168–9, 353
  setting, measuring, marking out and testing 41–2, 200–4
  shaping 43, 205–10
  smart and modern materials 29, 152, 170–2
  testing 36–7, 52
  see also wood
matrix 184, 188
measuring
  importance of 200
  and marking out 41, 201, 204
  precision instruments 203
mechanical advantage 265, 266–7, 272, 277–9
mechanical properties, materials 251, 252

mechanisms
  definition 265, 272
  linkages 268, 269, 272
  load, effort and fulcrum 266–8, 272
  mechanical advantage, velocity ratio and efficiency 265–6, 266–7, 269–71, 272, 277–9, 282–4, 285, 286–7
  motion see motion
  pulleys 269–70, 285, 286–8
  selecting for design applications 291
medium density fibreboard (MDF) 26, 46, 175, 180, 224
meganewton (MN) 255
melamine formaldehyde (MF) 175
members see structural members
memory, materials 170, 172
metals
  alloys 186, 189, 190, 191–2, 195
  changing structure and properties of 192–4
  extraction 189
  ferrous and non-ferrous 189, 190–1, 195
  finishes 43, 222, 223
  joining 44, 45, 211–14, 217, 218, 256–7
  properties 189–90, 251–2
  rendering 141
  shape memory alloys (SMA) 152, 170, 171
  shaping 206–7, 209
meters 303–4, 307
micro-switch 311, 315
mild steel 190, 192, 194, 212
mind map 19, 25
mock-ups, media for 26–9, 33
modelling 150–1
  CAD/CAM 27, 28, 99, 100, 152
  construction kits 28, 37
  generating ideas 24
  handmade models 99
  importance of 99, 150–1
  joining methods 152–5
  materials 26–7, 151–2, 168
  mock-ups, media for 26–9, 33
  Project work 346
  scale model 26, 29, 99, 150, 151, 155
  sketch model 26, 29
  software 24, 26, 33
  testing proposals with 97–8, 99
  tools 27, 99

moments 248–52
moments of force 248, 251, 252
monostable output 331, 332–3, 334
mortise and tenon joint 202, 215, 221
motion
   conversion of 273–80
   transmission of 281–91
   types of 273
moulding 43, 162, 173, 174, 176, 210
moulds 65, 68, 80
multimeters 303–4, 307

N

nails 44, 218–19
NAND gate 333, 335, 336, 337, 338, 339
nanomaterials 172
nanoparticles 170, 172
natural structures 235–6
Newson, Marc 93, 96
Newton 247, 255
nominal resistance value 316, 319, 327
non-ferrous metals 189, 191, 195
non-polarised capacitors 329
NOR gate 335
normally open/normally closed switches 312–13, 315
NOT gate 335
NPN transistors 320–1, 322
nuts and bolts 44, 45, 217, 218, 255
nylon 166, 174, 218

O

oak 179, 183, 209
off-shoring 60, 62
offset lithography printing 148
Ohm's Law 300
one-off production 64, 80
one-point perspective drawings 120, 132
open-loop systems 231–2
operation 230–1
operational amplifiers 339–43
OR gate 329, 333, 335, 336
orthographic projection 23, 25, 33, 110–11, 121
   laying out 136–7
   marking out 41
   Project work 355
   scale models 151
oscillating motion 273, 280
out-sourcing 60, 62
outcome 8, 10
output 69–71, 230–1

P

packaging 127
   blister packaging 160, 162, 163
   developments/nets 64, 68, 127–8, 129, 162
   evaluating 49–50
   recyclable 104–5
packaging net 64, 68
paint 43, 223, 224, 225
Panton, Verner 92, 95
paper 26, 43, 151
paper trimmer 159
PAR (Planed All Round) timber 196, 199
parallelogram of forces method 262, 263–4
parts list, exploded drawing 118
pass-fail analysis 30
patterns 65, 68, 93
PCB circuit layout 64, 68
peer review 52–3
permanent joining 154–5, 213, 255, 256
perpendicular 197, 198, 199, 202
Personal Protective Equipment (PPE) 57–8, 62, 186, 225–6
perspective 33
   one-point perspective 120, 132
   two-point perspective 113, 114, 120, 137–8
phenol formaldehyde (PF) 175
physical properties of materials 167, 169, 251, 252
physiological study 76, 77
pie charts 145
pilot hole 208, 217
plain bearings 295, 296, 297
planes 197, 206
planned obsolescence 77, 105–6
planning
   importance of 86
   for manufacture 37–9, 353–5
   methods 87
planometric drawings 112, 114
plastic state 173, 176
plastics 173–6
   acetate sheets 152
   finishes 43, 225
   joining 44, 45
   recycling 106, 176
   rendering 141
   shaping 43, 206–7, 209
   see also polymers
plywood 180, 181
pneumatic systems 288–90, 291
point load 245–6, 247
point of sale display 64, 68
polarised capacitors 328, 329
polishes 43, 223, 224, 225
polyester resin 175, 185
polyethylene terephthalate (PET) 174, 176
polymers 172, 173
   carbon-fibre reinforced polymers (CFRP) 185, 187
   glass-fibre reinforced polymers (GRP) 185
   shape memory polymers 170, 171
   see also plastics
polymorph 26, 29, 152, 171
polypropylene (PP) 174, 176
polystyrene (PS) 174, 176
polyvinyl chloride (PVC) 66, 174, 176
Postmodernism 95, 96
potential difference 300, 304
potential divider circuit 318, 319, 326, 327
power, electrical 300–1
preparation 225, 226, 227
presentation 139–43, 351–2
presentation board 34–5
primary function 90, 91
primary research 14–15, 16, 17
printed circuit boards (PCBs) 305–8, 310
printer's marks 149
printing 44, 148–9
   3D printing 28, 65, 66, 100
printing plate 148, 149
process colours 148, 149
process or control 69–71
product evaluation 16, 17, 48–54, 84–5, 356–8
production *see* manufacturing
production systems 64–6, 80–2
programmable integrated circuits 333–4
programmable interface controller (PIC) integrated circuits 334
The Project
   guidance 344, 346–7
   stages of 347–58
prototype 150, 151, 155, 157, 159, 169
   rapid prototyping 28, 66, 68, 100

prototype board 306, 309, 310
protractors 134
'Ps' analysis 11–12, 349
psychological study 76–7
PTC Creo 23, 64
PTM (press to make) and PTB (press to break) switches 311, 314, 338
pulleys 269–70, 286–8, 291
PVA (Polyvinyl Acetate) adhesive 44, 182, 220

Q

qualitative data 14, 17
quality control (QC) checks 61, 63, 65
   computerised 67
quantitative data 14, 17
questionnaires 12, 15, 48, 52

R

rack and pinion systems 281, 285–6
radio-frequency identification (RFID) devices 67
Rams, Dieter 93, 96
random shapes 22
rapid prototyping 28, 66, 68, 100
Rashid, Karim 93, 96
ratchet and pawl 286
reaction forces 245–6, 247
rechargeable batteries 292–3, 294
reciprocating motion 273, 280
recycling 104–7, 176
reducing shapes 130–3
reed switches 312
reforming 205, 209, 210
regular polygons 122–3, 126, 130
reinforced concrete 185, 186, 233, 236, 242, 251, 252
reinforcement 184, 188
relays 302, 313–15
remote manufacturing 65, 68
rendering 25, 140, 143
   CAD and 23, 24, 33, 64
   materials and textures 141–2
renewable energy sources 102
reprographics 148–9
research
   design brief 348–50
   importance of 14, 30
   organising 14–15, 16
   product analysis 16, 17, 72–3
residual current device (RCD) 292
resin identification codes 106, 176

resistance 300, 304, 316
   measuring 303–4
resistant materials 210, 344, 346
resistor tolerance 316–17, 319
resistors 316–19, 324, 326, 327, 332–3, 338–9
resolution 148, 149, 156–7
resources for designers 19
resultant 263, 264
reverse bias 323, 325
risk assessment 56–7, 62
riveting 44, 212–14, 255, 257
roller bearings 295, 296, 297
rotary cutter 160
rotary motion 273, 280
   linear motion, conversion 265, 273–7, 280, 281, 285–6
rotary switches 312
rulers 42, 134, 197

S

safety rule 159
saws 197, 206, 207
scale 23, 25, 111, 150
scale drawings 97, 151
scale model 26, 29, 99, 150, 151, 155
SCAMPER (substitute, combine, adapt, modify, purpose, eliminate, rearrange) strategy 21–2, 84–5
score card 30
screen printing 148
screw jacks 277–8, 280
screw thread mechanisms 277–8
screwdrivers 193, 217
screws 44, 217–18, 255, 257
secondary function 90, 91
secondary research 15, 16, 17
sectional view 115–16, 119, 121
   part, revolved and removed sections 116–17
self-adhesive vinyl 152
self-finishing materials 222, 225, 227
semiconductors 320, 323, 324, 336, 343
sequence drawings 145–6
sequential instructions 39
set squares 134
setting, measuring, marking out and testing 41–2, 200–4
seven segment displays 324–5
shape memory alloys (SMA) 152, 170, 171

shape memory polymers 170, 171
shaping materials 43, 205–10
shaping tools 159–60
shear forces 260–1
shell structures 234–5, 236
Single Pole Single Throw (SPST) and Single Pole Double Throw (SPDT) switches 313
sketch model 26, 29
Sketch Up 23, 64
sketching *see* freehand drawing
skills 36, 40
smart and modern materials 29, 152, 170–2
smart finish 225, 227
social design issues 73
sociological study 76
software
   CAD software 23, 64, 157
   circuit design 305, 306–7, 330
   modelling 24, 26, 33
softwoods 177, 179, 183
solar cells 293, 294, 299
soldering 212, 256, 309, 310
space frames 241
special finish 227
   see also finishes
specification 12–13, 30, 73
   evaluating products against 50–1, 89, 356–7
   Project work 350–1
sprockets and chain drives 288–9
spur gears 284, 285
stainless steel 190, 251
standard components 45, 47
star diagrams 27, 31
Starck, Philippe 91, 93
steam bending 182, 209
stencils 134, 143
stents 170, 171
stiffness 169
stock control, computerised 67
stools, construction 221
strain 169, 258–9, 264
strain gauges 253–4, 327
stress 169, 259–60, 264
stress-strain graphs 261, 264
strip boards 308, 310
structural members 239
   nature of 242–4
   testing 253–4
structures
   definition 233
   joints in 255–7
   types of 233–6

struts 238, 239
    buckling 243
style 94
    design movements 95–6
styrofoam 152
substrate 148, 149
summative evaluation 48, 54
surface finish 194, 227
    see also finishes
surface plate 197, 203
surveys 14, 17
sustainability 10, 49
    designers and 73–4
    recycling 104–7
switches 311–15, 321, 338–9
SWOT (strengths, weaknesses, opportunities, threats)
    analysis 30
systems
    closed- and open-loop systems 231–2
    definition 69, 230
    Project work 344, 346
systems diagrams 69–71, 230–1

T

tables, design planning 87, 89
tabs
    developments/nets 128–9
    joining materials 153–5
tangential arcs 125, 126
tangents 125–6
target market 9, 10, 348
technical pens 134, 139
technologists 72, 96
technology
    in manufacturing 63–8
    systems 69–71
tempering 193
template 35, 42, 65, 80, 202
tension 255
testing
    materials 36–7, 52
    printed circuit boards 307
    product testing and evaluation 52–3, 356–8
    structural members 253–4
    using models 97–8, 99
thermistor 327, 338
thermochromic materials 152, 171
thermoforming plastics 160, 173–4, 176, 209
thermosetting plastics 173, 175, 176
thick-and thin-line technique 139–40

third-angle projection 110, 112, 114
thrust bearings 295, 297
ties 238, 239
timber 177, 183
    joining 44, 182, 211, 215–17, 217–19, 221, 255
    seasoning 178, 183
    shaping 182, 209
    see also wood
time delay circuits 330–4
tin 186, 191, 192
tolerance 46, 47
    resistor tolerance 316–17, 319
tool steel 193, 194, 210
tools
    creasing and folding tools 160
    cutting tools 43, 159–60, 162, 197, 206
    die cutting/die stamping machines 43, 127, 160, 163
    drilling 206, 207, 208
    instruments and drafting aids 134–5, 197
    for measuring, setting and marking out 41–2, 201–3
    metals in 193
    for modelling 27, 99
    power tools 197, 207
torque 283–4, 286–7
torsional forces 260
toughness 167, 190
tracking 142, 143
traffic lights appraisal 31
trammel method 124–5
transducers 326–7
transistors 320–2, 326
    see also integrated circuits; logic gates
triangulation 235, 240, 241
    of forces 262–3
truncation 128, 129
trusses 241
try square 42, 197, 202
TTL (Transistor Transistor Logic) logic gates 336
two-point perspective drawings 113, 114, 120, 137–8
typography 142, 143

U

uniformly distributed load 246–7
urea formaldehyde (UF) 175
user group 9, 10, 14–15, 97, 98
user trial 52, 54
users see end user

V

vacuum forming 43, 160–1, 163, 173, 209
vanishing point, drawing 113, 114, 120
variable resistors 318, 319, 338
varnish 224, 225
vector graphics 156, 157, 158, 262–4
velocity ratio 265–6, 269–70, 272, 277–9
    belt drives 286–7
    gears 270–1, 282–4, 285
veneer 179, 180–1, 182, 209
vinyl cutters 66
vocabulary, technical 25
voltage 299–300, 304
    measuring 303–4
voltage gain 340–1

W

warping 178, 183
wastage 43, 205, 208, 209, 210
welding 44, 212, 255, 256
Wheatstone bridge 252, 327
wheel and axle 278–9, 280
wood
    environment and 182–3
    finishes 43, 222, 224, 353
    joining timber 44, 182, 211, 215–17, 217–19, 221, 255
    manufactured boards 180–1, 183
    natural timber 177
    PAR (Planed All Round) timber 196, 199
    properties 251–2
    rendering 141–2
    seasoning 178, 183
    shaping timber 182, 206–7, 209
    types of timber 177, 179, 183
    wood joints 202, 215, 221
work hardening 190, 192
working drawings 23, 33, 352–3
working properties of materials 167, 169
worm and worm-wheel systems 284–5

X

x-axis 144, 146

Y

y-axis 144, 146
yield stress 261, 262

Z

zinc 191, 223

Acknowledgements

The publishers wish to thank the following for permission to reproduce photographs. Every effort has been made to trace copyright holders and to obtain their permission for the use of copyright materials. The publishers will gladly receive any information enabling them to rectify any error or omission at the first opportunity.

(t = top, c = centre, b = bottom, r = right, l = left)

Thanks to the students of Jerudong International School, Brunei, for contributing examples of their coursework on p17, p34, p97bl, p97c, p97bl, p97br, p98l, p98r, p349, p350, p351, p352, p353, p354, p355, p356, p357, p358.

Thanks also to Terry Bream for his photographs on p234b, p235l, p235r, p244, p307t, p307b, p308, p309b, p311b, p313t, p313b, p324b, p325, p342.

PHOTO ACKNOWLEDGEMENTS

P6–7 sayhmog/ Shutterstock, p9 bikeriderlondon/ Shutterstock, p10 Gemenacom/ Shutterstock, p11 WhyMePhoto/ Shutterstock, p17 Marcos Mesa Sam Wordley/ Shutterstock, p20 ArtWell/ Shutterstock, p21tcr tchara/ Shutterstock, p21br Harish Marnad/ Shutterstock, p21tr Riccardo Mayer/ Shutterstock, p21tl Golubovy/ Shutterstock, p21bl reach/ Shutterstock, p21bcl Lucie Lang/ Shutterstock, p21bcr aastock/ Shutterstock, p21cr Valentin Valkov/ Shutterstock, p21cl Evgeny Karandaev/ Shutterstock, p21tcl Photographee.eu/ Shutterstock, p21c CTR Photos/ Shutterstock, p28l michelangeloop/ shutterstock.com, p28r bilciu/ shutterstock.com, p30 Dusit/ Shutterstock, p31 oorka/ Shutterstock, p32 Slaven/ Shutterstock, p33 grafvision/ Shutterstock, p37 PathomP/ Shutterstock, p39 Kozachenko Maksym/ Shutterstock, p45t iceink/ Shutterstock, p45b kilukilu/ Shutterstock, p52 Ken Schulze/ Shutterstock, p54 Kheng Guan Toh/ Shutterstock, p58 Barry Barnes/ Shutterstock, p59tl Attl Tibor/ Shutterstock, p59tc stoonn/ Shutterstock, p59tr Stephen Marques/ Shutterstock, p59bl Photoonlife/ Shutterstock, p59bcl veronchick84/ Shutterstock, p59bcr veronchick84/ Shutterstock, p59br Bakhur Nick/ Shutterstock, p66t Matthias Pahl/ Shutterstock, p66c jurra8/ Shutterstock, p66b Christian Delbert/ Shutterstock, p67t Anatoly Vartanov/ Shutterstock, p67b abeadev/ Shutterstock, p76l Marc McRaw/ Shutterstock, p76c Piotr Marcinski/ Shutterstock, p76r artproem/ Shutterstock, p78 Sebastian Kaulitzki/ Shutterstock, p79tl ChiccoDodiFC/ Shutterstock, p79cl Naypong/ Shutterstock, p79br gdvcom/ Shutterstock, p79bl Andrey_Popov/ Shutterstock, p79c Bastian Weltjen/ Shutterstock, p79tr ID1974/ Shutterstock, p79cr zimmytws/ Shutterstock, p82l Volodymyr Krasyuk/ Shutterstock, p82r Bohbeh/ Shutterstock, p83tl Tashsat/ Shutterstock, p83cl danielo/ Shutterstock, p83tr Incomible/ Shutterstock, p83tc freesoulproduction/ Shutterstock, p83cr Thierry Pirsoul/ Shutterstock, p83bl klempa/ Shutterstock, p83bc konahinab/ Shutterstock, p83br Vasilius/ Shutterstock, p84tl puttography/ Shutterstock, p84tc Paul Looyen/ Shutterstock, p84bl Meoita/ Shutterstock, p84bc Stocksnapper/ Shutterstock, p84tr Kletr/ Shutterstock, p84r Celso Diniz/ Shutterstock, p84b ShaunWilkinson/ Shutterstock, p90t graphixmania/ Shutterstock, p90b StudioSmart/ Shutterstock, p91l Dinu's/ Shutterstock, p91cl Brian A Jackson/ Shutterstock, p91cr Moustache Girl/ Shutterstock, p91r Ewais/ Shutterstock, p91b Panom Pensawang/ Shutterstock, p92t kurtcan/ Shutterstock, p92b PremiumVector/ Shutterstock, p92c chromatos/ Shutterstock, p93t america365/ Shutterstock, p93ct WitchEra/ Shutterstock, p93cb tanatat/ Shutterstock, p93b Boontang Pusdee/ Shutterstock, p94t Everything/ Shutterstock, p94b fckncg/ Shutterstock, p95t Gorbash Varvara/ Shutterstock, p95ct lynnette/ Shutterstock, p95c DutchScenery/ Shutterstock, p95cb Claudio Divizia/ Shutterstock.com, p95b borisovv/ Shutterstock, p96t MagMac83/ Shutterstock, p96b Peter Kotoff/ Shutterstock, p97 Sergey Skomorokhov/ Shutterstock, p99 Peter Aleksandrov/ Shutterstock, p100l YAKOBCHUK VASYL/ Shutterstock, p100tr YAKOBCHUK VASYL/ Shutterstock, p100b Ozgur Guvenc/ Shutterstock, p102t Meryll/ Shutterstock, p102c turtix/ Shutterstock, p102b Alexey Kamenskiy/ Shutterstock, p104 Photographee.eu/ Shutterstock, p105 tele52/ Shutterstock, p106 NEGOVURA/ Shutterstock, p108–9 VLADGRIN/Shutterstock, p121 Sylvie Bouchard/Shutterstock, p128 Hurst Photo/ Shutterstock, p134t Odua Images/Shutterstock, p134ct Taigi/Shutterstock, p134c joingate/Shutterstock, p134cb Ilja Generalov/Shutterstock, p134b Lichtmeister/Shutterstock, p148t Moreno Soppelsa/

Shutterstock, p148b Petr Lerch/Shutterstock, p149 Dja65/Shutterstock, p151 Aykut Erdogdu/Shutterstock, p152t MilsiArt/Shutterstock, p152c ppl/Shutterstock, p152b Chatchai Somwat/Shutterstock, p157 Pixaby, p159 Lasse Kristensen/Shutterstock, p160t Olena Zaskochenko/Shutterstock, p160b Denis Semenchenko/Shutterstock, p164–5 millann/ Shutterstock, p170tl VladisChern/ Shutterstock, p170tr Istomina Olena/ Shutterstock, p170br hywards/ Shutterstock, p173 ollirg/ Shutterstock, p176 NoPainNoGain/ Shutterstock, p180t günther pichler/ Shutterstock, p180c Pavel Hlystov/ Shutterstock, p180b eyal granith/ Shutterstock, p181 Noppharat46/ Shutterstock, p182 dandesign86/ Shutterstock, p184 Vadim Ratnikov/ Shutterstock, p185t fabiodevilla/ Shutterstock, p185c woodygraphs/ Shutterstock, p185b Digital Storm/ Shutterstock, p186t Nadezda Murmakova/ Shutterstock, p186b Toa55/ Shutterstock, p187 sportpoint/ Shutterstock, p188 Aumm graphixphoto/ Shutterstock, p196 Kobets Dmitry/ Shutterstock, p197 Ozgur Coskun/ Shutterstock, p201tcr graja/ Shutterstock, p201tr Edward Westmacott/ Shutterstock, p201bl terekhov igor/ Shutterstock, p201bcl Bill McKelvie/ Shutterstock, p201bcr RedDaxLuma/ Shutterstock, p202tl Fouad A. Saad/ Shutterstock, p202tr FoodStocker/ Shutterstock, p202bl micha_h/ Shutterstock, p202br moprea/ Shutterstock, p204l Neamov/ Shutterstock, p204cl AlexGreenArt/ Shutterstock, p204cr Omegafoto/ Shutterstock, p204r jambro/ Shutterstock, p205tl Tischenko Irina/ Shutterstock, p205tc luckyraccoon/ Shutterstock, p205tr albund/ Shutterstock, p205cl RedDaxLuma/ Shutterstock, p205c Oliver Wilde Shutterstock, p205cr mikeledray/ Shutterstock, p205bl gresei/ Shutterstock, p205bc arkivanov/ Shutterstock, p205br Paolo Cremonesi/ Shutterstock, p206tl Jim Hughes/ Shutterstock, p206tc Anan Kaewkhammul/ Shutterstock, p206tr Olivier Le Moal/ Shutterstock, p206cl MNI/ Shutterstock, p206c tcsaba/ Shutterstock, p206cr Sean van Tonder/ Shutterstock, p206bl Coprid/ Shutterstock, p206bc Alexander Tolstykh/ Shutterstock, p210 STILLFX/ Shutterstock, p211l Praphan Jampala/ Shutterstock, p211r SasinT/ Shutterstock, p212 FooTToo/ Shutterstock, p218t Steve Cordory/ Shutterstock, p218r AdrianFinlay/ Shutterstock, p218l Thissatan Kotirat/ Shutterstock, p218c Tom Gowanlock/ Shutterstock, p219t Dmitrij Skorobogatov/ Shutterstock, p219b Zoltan Fabian/ Shutterstock, p222l NanD_PhanuwatTH/ Shutterstock, p222c chungking/ Shutterstock, p222r Denis Kovin/ Shutterstock, p225 Digital Storm/ Shutterstock, p228–9 Sailorr/ Shutterstock, p230 DUSAN ZIDAR/ Shutterstock, p233 Cedric Weber/ Shutterstock, p234t Hywit Dimyadi/ Shutterstock, p234c Divin Serhiy/ Shutterstock.com, p236l Natchapon L./ Shutterstock.com, p236r Baloncici/ Shutterstock, p237 Andrei Seleznev/ Shutterstock, p240 Padmayogini/ Shutterstock.com, p241 LovePHY/ Shutterstock, p246 Muuraha/ Shutterstock, p248 Kzenon/ Shutterstock, p258 blurAZ/ Shutterstock, p265 fotum/ Shutterstock, p284t wayfarerlife/ Shutterstock, p284tc Aerodim/ Shutterstock, p284bc Gwoeii/ Shutterstock, p284b ra3rn/ Shutterstock, p288 enterphoto/ Shutterstock, p290 Antonsov85/ Shutterstock, p292t Africa Studio/ Shutterstock, p292b Kiya Grafica/ Shutterstock, p293t nrqemi/ Shutterstock, p293b estike/ Shutterstock, p295t TUM2282/ Shutterstock, p295tc loracreative/ Shutterstock, p295bc safakcakir/ Shutterstock, p295b Alleksander/ Shutterstock, p303 Viktorija Reuta/ Shutterstock, p309t Ziga Cetrtic/ Shutterstock, p311tl Scott Rothstein/ Shutterstock, p311tr Pabkov/ Shutterstock, p316 Francisco Javier Gil/ Shutterstock, p319 DeSerg/ Shutterstock, p320 vlabo/ Shutterstock, p324t Tsekhmister/ Shutterstock, p326 Timothy Hodgkinson/ Shutterstock, p328 yurazaga/ Shutterstock, p331 suthiphong yina/ Shutterstock, p344–5 Brian A Jackson/ Shutterstock.

We would like to thank the following organisations for kind permission to reproduce logos and photographs:

p103 with permission from One Laptop per Child, p107 Practical Action logo with permission from Practical Action.

Authors

Justin Harris is Head of Design & Technology at Jerudong International School, Brunei and has been in post for over 10 years. He moved into international education after training and working in the UK. He is committed to raising the profile and standards of Design & Technology in educational institutions and has hosted a number of international conferences in recent years. He is recognised for his work to promote the international development and support of Design Education and has received an award from the Design and Technology Association (UK) for his contributions to this area.

Dawne Bell is Assistant Head of Secondary Education at Edge Hill University (UK) where she holds responsibility for Teaching, Learning and Assessment for all Secondary (11–19) Initial Teacher Education programmes. Prior to this Dawne led PGCE and Undergraduate Teacher Education courses in both Design & Technology, and Creative Art, Design & Technology. Dawne worked for over 15 years in schools, and was Head of the Faculty of Art, Design & Technology at a large secondary school in Merseyside. She is well published and writes within the context of school based STEM, and has also written more widely on Gender, Social Justice and Educational Technology.

Chris Hughes has wide experience in the industrial and education sectors. He has worked in the textile, mechanical and materials technology fields, as well as the chemical and electronics industrial sectors. Chris has taught Design & Technology in a number of UK secondary, post-sixteen and higher educational establishments. This experience includes senior lecturing positions in Design & Technology at Manchester Metropolitan University and Edge Hill University. He is currently an associate lecturer with the Open University. Chris has been involved with a number of educational initiatives in Europe and Asia which help provide an international perspective to work he does.

Matt McLain is a teacher educator and Head of Secondary Programmes at Liverpool John Moores University (UK), having previously led the PGCE in Design & Technology. He taught for 11 years in Greater Merseyside as an advanced skills teacher and lead practitioner. Matt is an active member of both the D&T Association and the D&T Expert Subject Advisory Group in England. He has recently been involved with developments in the National Curriculum and the subject content for GCSE and A Level. His wider research interests are the teacher values and teacher modelling in Design & Technology and sociotechnological human activity.

Stewart Ross has been teaching graphic and product design for 8 years and is currently teaching design at an international school in Hong Kong. He has a Master of Arts in Education, a first class degree in Graphic Design and is an assistant examiner for the International Baccalaureate Organization (IB). Stewart is particularly interested in the Makerspace movement and its impact on the Design & Technology curriculum, and has given talks on developments in this area for the 21st Century Learning Conference in Hong Kong.

David Wooff is a Senior Lecturer and Course Leader at Edge Hill University (UK), where he runs the BSc (Hons) Secondary Design & Technology with Qualified Teacher Status (QTS) degree course. Prior to moving to Edge Hill, David worked in schools in the North West of England for over a decade, holding both curriculum and pastoral roles. David is well published in a variety of areas within Design & Technology, and more widely in the areas of Technology Enhanced Learning and the use of Augmented Reality to support learning in the classroom.

Notes

Notes

Notes

Notes

Notes

DATE DUE